Cognitive Technologies

Managing Editors: D.M. Gabbay   J. Siekmann

Editorial Board: A. Bundy   J.G. Carbonell
M. Pinkal   H. Uszkoreit   M. Veloso   W. Wahlster
M.J. Wooldridge

For further volumes:
www.springer.com/series/5216

Stanislav Živný

The Complexity
of Valued Constraint
Satisfaction
Problems

 Springer

Stanislav Živný
University College
University of Oxford
Oxford, UK

Managing Editors
Prof. Dov M. Gabbay
Augustus De Morgan Professor of Logic
Department of Computer Science
King's College London
London, UK

Prof. Dr. Jörg Siekmann
Forschungsbereich Deduktions- und
Multiagentensysteme
DFKI
Saarbrücken, Germany

ISSN 1611-2482   Cognitive Technologies
ISBN 978-3-642-43456-3   ISBN 978-3-642-33974-5 (eBook)
DOI 10.1007/978-3-642-33974-5
Springer Heidelberg New York Dordrecht London

ACM Computing Classification (1998): F.2.2, G.2.1, I.2

This book is dedicated to the memory of my mother, Marta Živná (1954–2010)

Preface

*Computer science is no more about computers
than astronomy is about telescopes.*
Edsger Dijkstra

The topic of this book is the following optimisation problem: given a set of discrete variables and a set of functions, each depending on a subset of the variables, minimise the sum of the functions over all variables. This fundamental research problem has been studied within several different contexts of computer science and artificial intelligence under different names: Min-Sum Problems, inference in Markov Random Fields (MRFs) and Conditional Random Fields (CRFs), Gibbs energy minimisation, valued constraint satisfaction problems (VCSPs), and (for two-state variables) pseudo-Boolean optimisation. We present general techniques for analysing the structure of such functions and the computational complexity of the minimisation problem.

This book could not have been written without the support of Oxford's University College, which funded me through a Stipendiary Junior Research Fellowship in Mathematical and Physical Sciences for 3 years.

Many results in this book are joint work with Dave Cohen and Pete Jeavons, from whom I have learnt the ropes of academic work. Pete has also served as my mentor and Ph.D. supervisor at Oxford. I am grateful to both for their advice, support, and friendship. Some results from Chap. 2 are joint work with Páidí Creed. Some results described in Chap. 3 are joint work with Bruno Zanuttini. The results described in Chap. 7 were obtained in collaboration with Vladimir Kolmogorov, whom I met at the Tractability Workshop in Microsoft Research Cambridge in 2010, when I was a research intern there. The results presented in Chap. 8 are joint work with Johan Thapper, whom I met at the Algebraic CSP Workshop at the Fields Institute for Research in Mathematical Sciences. The Fields Institute kindly funded my attendance at the workshop. The UK Engineering and Physical Sciences Research Council (EPSRC), the Royal Society (RS), and the French National Research Agency (ANR) financed my trips to the University of Toulouse III. Chapter 9 briefly summarises some of the results that have come out of these very productive research visits to

Toulouse and have given me the opportunity to work with and learn from Martin Cooper. I am grateful to all the above-mentioned collaborators for the time spent together, fun we had, and everything I have learnt from them.

I am grateful to my sister, Radka, and my parents-in-law for their support. Last but not least, I would like to express my gratitude to my wonderful wife, Biying, for her love, encouragement, and inspiration.

Oxford, UK   Stanislav Živný
July 2012

Contents

Introduction

We can only see a short distance ahead,
but we can see plenty there that needs to be done.
Alan Turing

The main topic of this book is the following optimisation problem: given a set of discrete variables and a set of functions, each depending on a subset of the variables, minimise the sum of the functions over all variables. This fundamental research problem has been studied within several different contexts of artificial intelligence, computer science, and combinatorial optimisation under different names: Min-Sum Problems [268], MAP inference in Markov Random Fields (MRFs) and Conditional Random Fields (CRFs) [203, 267], Gibbs energy minimisation [128], valued constraint satisfaction problems [97], and (for two-state variables) pseudo-Boolean optimisation [34, 84].

We start off with several examples of interesting and well-studied problems that can be modelled in this way.

Example 1 (Satisfiability) The standard propositional SATISFIABILITY problem for ternary clauses, 3-SAT [125], consists in determining whether it is possible to satisfy a Boolean formula given as a conjunction of ternary clauses, where each clause is a set of three literals and each literal is either a variable or the negation of a variable. A generalisation of 3-SAT, the 3-MAX-SAT problem [125] consists in finding an assignment of 0s and 1s (representing FALSE and TRUE, respectively) to all variables in the given formula such that the number of satisfied clauses is maximised, or equivalently (with respect to exact solvability), finding an assignment of 0s and 1s to all variables such that the number of unsatisfied clauses is minimised.

Any 3-SAT instance can be easily seen as a minimisation problem with the objective function given by a sum of ternary $\{0, \infty\}$-valued functions, one function for each clause. For instance, clause $(x \vee \neg v \vee z)$ yields a ternary function f defined by $f(0, 1, 0) = \infty$ and $f(x, y, z) = 0$ otherwise.

Similarly, any 3-MAX-SAT instance can be easily seen as a minimisation problem with the objective function given by a sum of ternary $\{0, 1\}$-valued functions, one function for each clause. For instance, clause $(x \vee \neg v \vee z)$ yields a ternary function f defined by $f(0, 1, 0) = 1$ and $f(x, y, z) = 0$ otherwise.

Example 2 (Graph colouring) The k-COLOURABILITY problem [125] consists in determining whether it is possible to assign k colours to the vertices of a given graph so that adjacent vertices are assigned different colours. This can be viewed as a minimisation problem with the objective function given by a sum of binary functions; each edge in the graph yields a binary function f defined by $f(x, y) = 0$ if $x \neq y$ and $f(x, y) = \infty$ otherwise.

Example 3 (Digraph acyclicity) Given a directed graph G, the question of whether G is acyclic can be modelled as follows: variables correspond to the vertices of G, the domain of every variable is the set of natural numbers $\mathbb{N}$,[1] and every arc (x, y) of G yields a binary function f that represents the standard "smaller than" ordering on natural numbers; that is, $f(x, y) = 0$ if $x < y$ and $f(x, y) = \infty$ otherwise.

Example 4 (Diophantine equations) Hilbert's tenth problem asks for an algorithm that decides whether a given system of polynomial equations with integer coefficients (a *diophantine* equation system) has an integer solution. This problem can be modelled with variables $x_1, \ldots, x_n$, each with the domain $\mathbb{Z}$, and constraints of the form $ax_i + bx_j + c = x_k$, or $x_i * x_j = x_k$, for $a, b, c, d \in \mathbb{Z}$ and $i, j, k \in \{1, 2, \ldots, n\}$; this can be easily represented by $\{0, \infty\}$-valued ternary functions. Matiyasevič has shown that this problem is undecidable [216].

Example 5 (Min-Cost-Hom) Given a graph G, we denote by $V(G)$ the set of vertices of G and by $E(G)$ the set of edges of G. Given two (directed or undirected) graphs G and H, a mapping $h : V(G) \to V(H)$ is a *homomorphism* from G to H if h preserves edges; that is, $(u, v) \in E(G)$ implies $(h(u), h(v)) \in E(H)$. The homomorphism problem for graphs asks for the existence of a homomorphism from G to H [150]. Let $c_v(u)$ be a nonnegative rational cost for all $u \in V(G)$ and $v \in V(H)$. The cost of a homomorphism f from G to H is defined by $\sum_{u \in V(G)} c_{f(u)}(u)$. The MINIMUM-COST HOMOMORPHISM problem, MIN-COST-HOM [143, 145], asks for a homomorphism of minimum cost between two given graphs. This problem can be cast as a minimisation problem of a sum of binary $\{0, \infty\}$-valued functions and unary rational-valued functions.

Example 6 (Max-Cut) Given a graph G with the vertex set V, the MAXIMUM CUT problem, MAX-CUT [125], consists in finding a subset $S \subseteq V$ of the vertices of G that maximises the number of edges between vertices in S and $V \setminus S$. The polynomial-time equivalent (with respect to exact solvability) problem is MINIMUM UNCUT, MIN-UNCUT; that is, the problem of finding a subset $S \subseteq V$ that minimises the number of edges in S and $V \setminus S$.

   This problem can be seen as a minimisation problem with the objective function being a sum of binary functions defined on $\{0, 1\}$. In particular, if we define

---

[1]In fact, only the set $\{1, 2, \ldots, n\}$, where n is the number of vertices of G, would suffice as the domain of each variable.

$\bar{\lambda}(x, y) = 0$ if $x \neq y$ and $\bar{\lambda}(x, y) = 1$ if $x = y$, then the objective function is a sum of $\bar{\lambda}$'s, each edge of G corresponding to one $\bar{\lambda}$.

Example 7 ((s,t)-Min-Cut) Given a graph G with the vertex set V and two specified vertices $s, t \in V$, the (s, t)-MIN-CUT problem consists in finding a subset $S \subseteq V$ of the vertices G with $s \in V$ and $t \notin V$ that minimises the number of edges between vertices in S and $V \setminus S$.

This problem can be seen as a minimisation problem with the objective function being a sum of unary and binary functions defined on $\{0, 1\}$. In particular, let $\lambda(0, 1) = 1$ and $\lambda(x, y) = 0$ otherwise, and for any $d \in D$, let $\mu^d(d) = \infty$ and $\mu^d(x) = 0$ otherwise. Now the objective function is a sum of λ's, each edge of G corresponding to one λ, and $\mu^1(s) + \mu^0(t)$, which enforces the inclusion of s and exclusion of t.

It is straightforward to generalise (s, t)-MIN-CUT to graphs with edge weights. Both versions of the problem are solvable in polynomial time [131].

Example 8 (Submodularity) Let $D = \{1, 2, \ldots, d\}$ for some fixed d. Let Γ be a set of functions $f : D^k \to \mathbb{Q}_{\geq 0}$ satisfying $f(\min(s, t)) + f(\max(s, t)) \leq f(s) + f(t)$ for any $s, t \in D^k$, where k is the arity of f, and min and max are binary functions returning the smaller and larger, respectively, of its two arguments with respect to the usual order on integers. (Note that min and max are applied componentwise on tuples s and t.) Any sum of functions from Γ over n variables, each with domain D, can be minimised in polynomial time in n due to its submodularity property [159, 257].

We finish our list with some more applied examples.

Example 9 (Timetabling) In timetabling exams at a university [248], variables can represent the times and locations of the different exams, and the functions can model the capacity of each examination room (for example, we cannot assign more students to take exams in a given room at any one time than the room's capacity) and prevent certain exams from being scheduled at the same time (for example, we cannot schedule two exams at the same time if they share students in common). The objective function can also take into account teachers' and students' preferences.

Example 10 (Texture-based segmentation) Given a set of distinct textures, such as a dictionary of RGB patches, together with their object class labels, the goal is to segment a given image; that is, the pixels of the image should be labelled as belonging to one of the object classes. This problem can be formulated using discrete variables, one for each pixel, where the domain of each variable is the set of distinct object classes. The binary functions are usually defined such that they encourage contiguous segments whose boundaries lie on image edges [37]. Similar approaches can be used for 3D reconstruction or object recognition [36].

Example 11 (Office assignment) Each of n staff members, represented by n variables, must be assigned an office. There are m offices, each of which can be assigned at most u_j people. Unary functions express personal preferences of each staff member for each office. There are also nonoverlapping groups of people $G_1, \ldots, G_g$ whom we would prefer to assign to different offices (such as married couples, for example). What is the best solution?

While some of the problems from the examples above are tractable, such as Examples 7 and 8, some of them are (NP-)hard, or even undecidable, such as Example 4. *Our main interest is in the question of what makes these problems hard and what the special cases are that are tractable.*

Focus of This Book

The focus of this book is on *exact solvability*; that is, we are interested in solving the problem in hand *optimally* (as opposed to approximately). Furthermore, a class of problems is considered *tractable* if any instance from it can be solved in *polynomial time* (as opposed to other notions of tractability such as moderate exponential-time tractability or fixed-parameter tractability). Finally, we will consider problems with discrete variables on *finite domains* only. For some (classes of) problems, the domains will be fixed; for others, the domains will be part of the input (and thus unbounded), but always finite.

Since all the above-mentioned frameworks are equivalent with respect to exact solvability and given that this book is based on several papers that talk about valued constraint satisfaction problems, we will use the terminology of *valued constraint satisfaction problems* (VCSPs) [97, 253].

A special case of VCSPs are so-called *constraint satisfaction problems* (CSPs), first identified in the seminal work of Montanari [218]. CSPs deal only with the feasibility (rather than the optimisation) problem, as is Examples 2 and 3, but we will pay some attention to them in this book as they have been studied in other contexts as well, such as homomorphisms between relational structures [111, 150] and conjunctive query evaluation [111, 136, 183, 251]. Apart from the tractability notion studied in this book, defined by polynomial-time solvability, there is an alternative and well-studied approach to solving CSPs in practice, which consists in interleaving a backtracking search with a series of heuristics and polynomial-time propagation, which significantly prune the exponential search space. We refer the reader interested in this research area, known as *constraint programming*, to the standard textbooks [3, 97, 248], the proceedings of the Annual International Conference on Principles and Practice of Constraint Programming (CP), and the website of the Association for Constraint Programming (ACP).[2]

---

[2]http://4c.ucc.ie/a4cp/.

Structure of This Book

Apart from the introduction, this book consists of three parts. Part II is based on the author's doctoral thesis from the University of Oxford [274], which won the 2011 Association for Constraint Programming (ACP) Doctoral Award.

Part II investigates the expressive power of various classes of functions. Chapter 2, based on [63, 68, 274], presents an algebraic theory for the expressive power of languages. Chapter 3, based on [70] (preliminary version [69]) and [272], investigates the expressive power of fixed-arity languages. Chapter 4, based on [280] (preliminary version [277]), is concerned with submodular languages. Chapter 5, based on [275] (preliminary version [276]), shows that not all submodular languages are expressible by binary submodular languages.

Part III deals with the tractability of valued constraint satisfaction problems. Chapter 6 surveys known tractable languages and is based on [168]. Chapter 7, based on [188] (full version [187]), briefly presents the complexity of conservative languages. Chapter 8, based on [264], presents recent results on the power of linear programming for valued constraint satisfaction problems. Chapter 9, based on [80] and [82] (preliminary versions [78, 79, 81]), surveys known results on hybrid tractability. Finally, Chap. 10 concludes with some open problems.

Part I
Background

Chapter 1
Background

The highest technique is to have no technique.
Bruce Lee

1.1 Introduction

In this chapter, we will introduce the necessary background. First, we will define valued constraint satisfaction problems and give examples of problems that can be formulated as such. Second, we will define the key concept of expressibility and introduce certain algebraic properties that play a key role in the complexity analysis of VCSPs. Finally, we will present the notion of submodularity and survey known results regarding optimising submodular functions.

1.2 Valued Constraint Satisfaction Problems

For a tuple t, we shall denote by $t[i]$ its ith component. We shall denote by $\mathbb{Q}$ the set of all rational numbers, by $\mathbb{Q}_{\geq 0}$ the set of all nonnegative rational numbers, and by $\overline{\mathbb{Q}}_{\geq 0}$ the set of all nonnegative rational numbers together with (positive) infinity, ∞. We define $\alpha + \infty = \infty + \alpha = \infty$ for all $\alpha \in \overline{\mathbb{Q}}_{\geq 0}$ and $\alpha\infty = \infty$ for all $\alpha \in \mathbb{Q}_{\geq 0}$. Members of $\overline{\mathbb{Q}}_{\geq 0}$ are called *costs*.

For any fixed set D, a function ϕ from D^m to $\overline{\mathbb{Q}}_{\geq 0}$ will be called a *cost function* on D of arity m. D is called a *domain*, and in this book we will only deal with finite domains. If the range of ϕ lies entirely within $\mathbb{Q}_{\geq 0}$, then ϕ is called a *finite-valued* cost function. If the range of ϕ is $\{0, \infty\}$, then ϕ is called a *crisp* cost function. If the range of a cost function ϕ includes both nonzero finite costs and infinity, we emphasise this fact by calling ϕ a *general-valued* cost function.

With any *relation* R on D we can associate a crisp cost function ϕ_R on D which maps tuples in R to 0 and tuples not in R to ∞. On the other hand, with any m-ary cost function ϕ we can associate a relation R_ϕ defined by $\langle x_1, \ldots, x_m \rangle \in R_\phi \Leftrightarrow \phi(x_1, \ldots, x_m) < \infty$, or equivalently an m-ary crisp cost function defined by:

$$\mathsf{Feas}(\phi)(x_1, \ldots, x_m) \stackrel{\text{def}}{=} \begin{cases} \infty & \text{if } \phi(x_1, \ldots, x_m) = \infty, \\ 0 & \text{if } \phi(x_1, \ldots, x_m) < \infty. \end{cases}$$

S. Živný, *The Complexity of Valued Constraint Satisfaction Problems*,
Cognitive Technologies, DOI 10.1007/978-3-642-33974-5_1,
© Springer-Verlag Berlin Heidelberg 2012

We call Feas($\cdot$) the *feasibility operator*. In view of the close correspondence between crisp cost functions and relations we shall use these terms interchangeably in the rest of the book.

A VCSP instance consists of a set of variables, a set of possible values, and a multiset of constraints. Each constraint has an associated cost function which assigns a cost (or a degree of violation) to every possible tuple of values for the variables in the scope of the constraint. The goal is to find an assignment of values to all of the variables that has the minimum total cost.

Definition 1.1 (VCSP) An instance $\mathcal{P}$ of the VALUED CONSTRAINT SATISFACTION problem, VCSP, is a triple $\langle V, D, \mathcal{C} \rangle$, where V is a finite set of *variables*, which are to be assigned values from the set D, and $\mathcal{C}$ is a multiset of *constraints*. Each $c \in \mathcal{C}$ is a pair $c = \langle \mathbf{x}, \phi \rangle$, where $\mathbf{x}$ is a tuple of variables of length m, called the *scope* of c, and $\phi : D^m \to \overline{\mathbb{Q}}_{\geq 0}$ is an m-ary cost function. An *assignment* for the instance $\mathcal{P}$ is a mapping s from V to D. We extend s to a mapping from V^k to D^k on tuples of variables by applying s componentwise. We denote by $\mathcal{A}$ the set of all assignments. The *cost* of an assignment s is defined as follows:

$$\mathsf{Cost}_{\mathcal{P}}(s) \stackrel{\text{def}}{=} \sum_{\langle \mathbf{x}, \phi \rangle \in \mathcal{C}} \phi\big(s(\mathbf{x})\big).$$

A *solution* to $\mathcal{P}$ is an assignment with minimum cost.

Definition 1.2 (CSP) An instance $\mathcal{P}$ of the CONSTRAINT SATISFACTION problem, CSP [248], is a VCSP instance where all cost functions are crisp, that is, relations. The task of finding an assignment with minimum cost amounts to testing whether all constraints can be satisfied (zero cost) or not (infinite cost).

Remark 1.1 Constraints are often called valued constraints in the VCSP literature to distinguish them from crisp constraints. We will use the terms "constraints" and "valued constraints" interchangeably, and use the term "crisp constraints" for constraints whose associated cost functions are relations.

Remark 1.2 In the original, more general, definition of the VCSP [21], costs were allowed to lie in any positive tomonoid S called a *valuation structure*.[1] Under the additional assumptions of discreteness and the existence of a partial inverse operation, it has been shown [71] that such a structure S can be decomposed into independent positive tomonoids, each of which is isomorphic to a subset of $\mathbb{R}_{\geq 0}$ with the operation being either standard addition, $+$, or bounded addition, $+_k$, where $a +_k b = \min(k, a+b)$. Therefore, using $\mathbb{R}_{\geq 0}$ instead of an arbitrary valuation structure, we do not restrict ourselves too much. Moreover, using costs from $\mathbb{R}_{\geq 0}$ and combining them using addition is standard in operational research.

---

[1] A *valuation structure*, Ω, is a totally-ordered set, with a minimum and a maximum element (denoted by 0 and by ∞), together with a commutative, associative binary *aggregation operator*, $\oplus$, such that for all $\alpha, \beta, \gamma \in \Omega$, $\alpha \oplus 0 = \alpha$ and $\alpha \oplus \gamma \geq \beta \oplus \gamma$ whenever $\alpha \geq \beta$.

In order to avoid difficulties with representation issues for transcendental numbers such as π or e, we restrict ourselves to nonnegative rationals. However, most tractability results mentioned in this book also hold for (nonnegative) reals with a suitable representation, such as algebraic reals, that is, reals that are roots of nonzero polynomials in one variable with rational (or equivalently, integer) coefficients.

Remark 1.3 Infinite costs can be used to indicate infeasible assignments (hard constraints), and hence the VCSP framework includes the standard CSP framework as a special case and is equivalent to the CONSTRAINT OPTIMISATION problem framework, COP, which is widely used in practice [248].

Remark 1.4 A number of extensions have been added to the basic CSP framework to deal with questions of optimisation, including semi-ring CSPs, valued CSPs, soft CSPs, and weighted CSPs. These extended frameworks can be used to model a wide range of discrete optimisation problems [21, 72, 161, 248, 253], including standard problems such as (s, t)-MIN-CUT, MAX-SAT, MAX-ONES SAT, MAX-CSP [67, 87], and MIN-COST HOMOMORPHISM [145].

The differences between the various general frameworks for soft CSPs, such as valued CSPs or semi-ring CSPs, are not relevant for our purposes. The semi-ring CSP framework is slightly more general,[2] but the valued CSP framework is sufficiently powerful to model a wide range of optimisation problems. Hence we will simply focus on this one very general framework, VCSP.

Remark 1.5 We remark on terminological differences. VCSPs are studied under different names such as Min-Sum, Gibbs energy minimisation, and Markov Random Fields; domain values are sometimes called *labels*, whereas binary instances are called *pairwise* instances, m-ary cost functions are called m-cliques, and assignments are called *labellings*.

We show now that many classical problems can be formulated as subproblems of the VCSP.

Example 1.1 (CSP) For any instance of the classical constraint satisfaction problem $\mathcal{P} = \langle V, D, \mathcal{C} \rangle$, we define a corresponding valued constraint satisfaction problem instance $\mathcal{P}' = \langle V, D, \mathcal{C}' \rangle$. For each constraint $\langle \sigma, R \rangle \in \mathcal{C}$ we define a cost function ϕ_R and set $\mathcal{C}' = \{\langle \sigma, \phi_R \rangle | \langle \sigma, R \rangle \in \mathcal{C}\}$. The cost function σ_R maps each tuple allowed by R to 0, and each tuple disallowed by R to ∞. Any solution s to $\mathcal{P}'$ has cost 0 if, and only if, s satisfies all the constraints of $\mathcal{P}$.

Example 1.2 (Boolean conjunctive query evaluation) It is well known that certain fundamental problems in database theory such as BOOLEAN CONJUNCTIVE

---

[2]The main difference between semi-ring CSPs and valued CSPs is that costs in valued CSPs represent violation levels and have to be totally ordered, whereas costs in semi-ring CSPs represent preferences and may be ordered only partially.

QUERY EVALUATION and CONJUNCTIVE QUERY CONTAINMENT are equivalent to the CSP [111, 117, 136, 183, 251], and hence can be formulated in the VCSP by Example 1.1.

Example 1.3 (Min-CSP/Max-CSP) An instance $\mathcal{P}$ of the MAXIMUM CONSTRAINT SATISFACTION problem [87] is an instance of the CSP with the goal of maximising the number of satisfied constraints. In the weighted version, each constraint has a nonnegative weight and the goal is to maximise the weighted number of satisfied constraints. We shall denote by MAX-CSP the more general weighted version.

Maximising the weighted number of satisfied constraints is the same as minimising the weighted number of unsatisfied constraints (MIN-CSP).[3] Hence for any instance $\mathcal{P}$ of the MAX-CSP or MIN-CSP, we can define a corresponding VCSP instance $\mathcal{P}'$ in which a constraint c with weight w is associated with a cost function that maps tuples allowed by c to 0 and tuples disallowed by c to w.

Example 1.4 (Max-Ones) MAX-ONES is an extension of the Boolean CSP framework in which the goal is to satisfy all given constraints and simultaneously maximise the number of variables assigned the value 1 [87]. In the weighted version each variable has a nonnegative weight and the goal is to maximise the weighted number of variables assigned the value 1. We shall denote by MAX-ONES the more general weighted version.

As in the MAX-CSP from Example 1.3, maximising the weighted number of variables assigned the value 1 is the same as minimising the weighted number of variables assigned the value 0. Hence for any instance $\mathcal{P}$ of MAX-ONES, we can define an instance $\mathcal{P}'$ of the VCSP that has the same variables, domain, and constraints as $\mathcal{P}$, with additional unary constraints: on a variable with weight w, we impose a unary constraint with the cost function μ defined by $\mu(0) = w$ and $\mu(1) = 0$.

Example 1.5 (Min-Ones) MIN-ONES is an extension of the Boolean CSP framework in which the goal is to satisfy all given constraints and simultaneously minimise the number of variables assigned the value 1 [87]. In the weighted version each variable has a nonnegative weight and the goal is to minimise the weighted number of variables assigned the value 1. We shall denote by MIN-ONES the more general weighted version.

For any instance $\mathcal{P}$ of MIN-ONES, we can define an instance $\mathcal{P}'$ of the VCSP that has the same variables, domain, and constraints as $\mathcal{P}$, with additional unary constraints: on a variable with weight w, we impose a unary constraint with the cost function μ defined by $\mu(0) = 0$ and $\mu(1) = w$.

Example 1.6 ((s,t)-Min-Cut) Let $G = \langle V, E \rangle$ be a directed weighted graph such that for every $(u, v) \in E$ there is a weight $w(u, v) \in \overline{\mathbb{Q}}_{\geq 0}$ and let $s, t \in V$ be the source and target nodes. An (s, t)-*cut* C is a subset of vertices V such

---

[3]This is true for optimal solutions. However, if we are interested in approximability results, this statement is not true even over Boolean domains; see [87].

that $s \in C$ but $t \notin C$. The weight, or the size, of an (s, t)-cut C is defined by $\sum_{(u,v)\in E, u\in C, v\notin C} w(u, v)$. The (s, t)-MIN-CUT problem consists in finding a minimum-weight (s, t)-cut in G.

We can formulate the search for a minimum-weight (s, t)-cut in G as a VCSP instance. For a fixed weight $w \in \overline{\mathbb{Q}}_{\geq 0}$, we define

$$\lambda_w(x, y) \stackrel{\text{def}}{=} \begin{cases} w & \text{if } x = 0 \text{ and } y = 1, \\ 0 & \text{otherwise.} \end{cases}$$

For a fixed value $d \in \{0, 1\}$ and a cost $w \in \overline{\mathbb{Q}}_{\geq 0}$, we define

$$\mu_w^d(x) \stackrel{\text{def}}{=} \begin{cases} w & \text{if } x = d, \\ 0 & \text{if } x \neq d. \end{cases}$$

We denote by Γ_{cut} the set of cost functions λ_w and μ_w^d.

Let $\mathcal{P} = \langle V, \{0, 1\}, \{\langle\langle u, v\rangle, \lambda_{w(u,v)}\rangle \mid (u, v) \in E\} \cup \{\langle s, \mu_\infty^1\rangle, \langle t, \mu_\infty^0\rangle\}\rangle$.

The unary constraints ensure that the source and target nodes take the values 0 and 1, respectively. Therefore, a minimum-weight (s, t)-cut in G corresponds to the set of variables assigned the value 0 in some solution to $\mathcal{P}$.

1.3 Complexity of VCSPs

For each VCSP instance there is a corresponding decision problem in which the question is to decide whether there is a solution with cost lower than some given threshold value. It is clear from Example 1.1 that there is a polynomial-time reduction from the CSP to this decision problem. Since the CSP is known to be NP-complete [206], it follows that the VCSP is NP-hard.

The problem of finding a solution to a VCSP instance is an NP optimisation problem; that is, it lies in the complexity class NPO. Informally, NPO consists of function problems of the form "find an assignment of the variables $x_1, \ldots, x_k$ that minimises a cost function $\phi(x_1, \ldots, x_k)$, where ϕ is computable in polynomial time"; see [6] for a formal definition of NPO. The VCSP framework is powerful enough to describe many NPO-complete problems, for instance, MAX-ONES from Example 1.4, also known as MAXIMUM WEIGHTED SATISFIABILITY [6].

One significant line of research on the VCSP is identifying restrictions that ensure that instances are solvable in polynomial time. There are two main types of restrictions that have been studied in the literature.

1.3.1 Structural Restrictions

First, we can limit the *structure* of the instances, in the following sense. With any instance $\mathcal{P}$ of the VCSP, we can associate a hypergraph $\mathcal{H}_{\mathcal{P}}$ whose vertices are the variables of $\mathcal{P}$, and whose hyperedges correspond to the scopes of the constraints of $\mathcal{P}$. The hypergraph $\mathcal{H}_\mathcal{P}$ is called the structure of $\mathcal{P}$, and is also known as the *constraint network* [97].

A number of results concerning restrictions to the structure of problem instances that are sufficient to ensure tractability have been obtained for the CSP framework, and can be easily generalised to the VCSP [17, 97, 132].

For example, if $\mathcal{H}_\mathcal{P}$ is "tree-like" in various ways, then it can be shown that $\mathcal{P}$ is solvable in polynomial time via dynamic programming, as first observed in the case of binary CSPs in the pioneering work [218]. This observation has since been generalised in many different ways [1, 56, 61, 93, 98, 119, 120, 133–135, 140, 146, 183, 212, 213, 215]. In general, the structural restrictions that ensure tractability are those that enforce a bound on some measure of width in the class of structures allowed [133]. Complete classifications, identifying all tractable cases, have been obtained for bounded-arity CSPs [139], and for unbounded-arity CSPs [214].

1.3.2 Language Restrictions

However, the complexity of finding an optimal solution to a VCSP instance will obviously also depend on the forms of constraints that are allowed in the problem [15, 67]. Restricting the *forms* of the constraints allowed in the problem gives rise to so-called *language restrictions*.

Definition 1.3 (Language) A *valued constraint language*, a *constraint language*, or just a *language*, is any set Γ of cost functions on some fixed set D.

Most work on language-restricted CSPs and VCSPs deals with languages on fixed domains. Hence, unless stated otherwise, we assume through the rest of this section and book that D is fixed. (The only exceptions will be the last part of this section and Chap. 9, which both deal with hybrid tractability.)

Notation 1.1 A language Γ over a two-element domain D is called *Boolean*.

Notation 1.2 A language Γ is called *crisp* if all cost functions from Γ are crisp. Similarly, Γ is called *finite-valued* (or *general-valued*) if all cost functions from Γ are finite-valued (or general-valued).

Notation 1.3 We will denote by VCSP(Γ) the class of all VCSP instances in which all cost functions in all constraints belong to Γ.

Example 1.7 Let $D = \{0, 1\}$. We define two unary cost functions as follows:

$$\mu_2(x) \overset{\text{def}}{=} \begin{cases} 0 & \text{if } x = 0, \\ 5 & \text{if } x = 1. \end{cases}$$

$$\mu_5(x) \stackrel{\text{def}}{=} \begin{cases} 4 & \text{if } x = 0, \\ 2 & \text{if } x = 1. \end{cases}$$

We also define six binary cost functions by the following table:

	ϕ_{12}	ϕ_{14}	ϕ_{23}	ϕ_{34}	ϕ_{35}	ϕ_{45}
00	3	0	0	9	3	4
01	2	4	1	7	5	3
10	3	2	0	8	4	2
11	1	5	0	1	4	1

The set $\Gamma = \{\mu_2, \mu_5, \phi_{12}, \phi_{14}, \phi_{23}, \phi_{34}, \phi_{35}, \phi_{45}\}$ is an example of a valued constraint language. We will now give an example of a VCSP(Γ) instance. Let $V = \{x_1, x_2, x_3, x_4, x_5\}$ be a set of variables, and let $\mathcal{C}$ be a set of constraints, defined by:

$$\mathcal{C} = \{\langle\langle x_1, x_2\rangle, \phi_{12}\rangle, \langle\langle x_1, x_4\rangle, \phi_{14}\rangle, \langle\langle x_2, x_3\rangle, \phi_{23}\rangle,$$

$$\langle\langle x_3, x_4\rangle, \phi_{34}\rangle, \langle\langle x_3, x_5\rangle, \phi_{35}\rangle, \langle\langle x_4, x_5\rangle, \phi_{45}\rangle, \langle x_2, \mu_2\rangle, \langle x_5, \mu_5\rangle\}.$$

Then $\mathcal{P} = \langle V, D, \mathcal{C}\rangle$ is a VCSP(Γ) instance, illustrated in Fig. 1.1.

Notation 1.4 A valued constraint language Γ is called *tractable* if, for every finite subset $\Gamma' \subseteq \Gamma$, there exists an algorithm solving any instance from VCSP(Γ') in polynomial time. Conversely, Γ is called *intractable*, or NP-hard, if VCSP(Γ') is NP-hard for some finite subset $\Gamma' \subseteq \Gamma$.

Remark 1.6 A tractable Γ is sometimes called *locally tractable* as opposed to *globally tractable*, where the latter means that there is a uniform polynomial-time algorithm that solves any instance from VCSP(Γ') for all finite subsets $\Gamma' \subseteq \Gamma$.

Remark 1.7 Defining tractability in terms of finite subsets ensures that the tractability of a valued constraint language is independent of whether the cost functions are represented *explicitly* (via tables of values) or *implicitly* (via oracles). This is because for any finite $\Gamma' \subseteq \Gamma$, the algorithm for solving VCSP(Γ') can remember all the values of all cost functions in Γ'.

Our interest is in the effect of restricting the forms of cost functions allowed in valued constraint languages. In some cases, the restriction on the valued constraints may result in more tractable versions of the VCSP.

Example 1.8 (Γ_{cut}) Recall the valued constraint language Γ_{cut} from Example 1.6. Any instance of VCSP(Γ_{cut}) on variables $x_1, \ldots, x_n$ can be solved in $O(n^3)$ time by a standard (s, t)-MIN-CUT algorithm [131], via the following reduction to (s, t)-MIN-CUT: any unary constraint μ_w^0 (or μ_w^1) on x_i can be modelled by an edge of weight w from x_i to the target node (or from the source node to the node x_i). Any $\lambda_w(x_i, x_j)$ constraint is modelled by an edge of weight w from x_i to x_j.

Fig. 1.1 Instance $\mathcal{P}$
(Example 1.7)

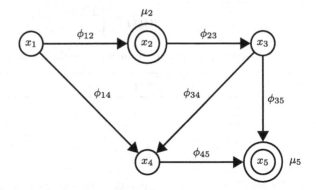

Example 1.9 (Min-Cut) Given a directed weighted graph $G = \langle V, E \rangle$ as in Example 1.6, a subset of vertices $C \subseteq V$ is called a *cut* if C is nontrivial, that is, $C \neq \emptyset$ and $C \neq V$. The weight of C is defined as in Example 1.6. The MIN-CUT problem, also known as the GLOBAL MIN-CUT problem, consists in finding a minimum-weight cut in G. Using the cubic-time algorithm for the (s, t)-MIN-CUT problem [131], one can easily construct an algorithm for the MIN-CUT problem of order $O(n^4)$, where $n = |V|$ is the number of vertices of G. For undirected graphs, Nagamochi and Ibaraki have described a more efficient algorithm, still based on network flows, which runs in cubic time [221]. Later, a simpler and purely combinatorial cubic-time algorithm not based on network flows was discovered [260].

It has been shown in [279] that MIN-CUT cannot be naturally described in the VCSP framework by any tractable valued constraint language over finite domains.

Next we give a list of several well-known problems that can be formulated in the language-restricted VCSP framework. We have already seen some examples earlier in this section and also in the introduction.

We refer to the original papers that provide complexity *classification* results. In other words, they show under which language restrictions a given problem is tractable and under which restrictions it is intractable.

We start with an example that demonstrates that a valued constraint satisfaction problem involving only one single binary Boolean cost function can be NP-hard.

Example 1.10 (Γ_{xor}) Let Γ_{xor} be the Boolean valued constraint language which contains just the single binary cost function $\phi_{\text{xor}} : D^2 \to \overline{\mathbb{Q}}_{\geq 0}$ defined by

$$\phi_{\text{xor}}(x, y) \overset{\text{def}}{=} \begin{cases} 1 & \text{if } x = y, \\ 0 & \text{if } x \neq y. \end{cases}$$

If $D = \{0, 1\}$, then the problem VCSP(Γ_{xor}) corresponds to the MAX-SAT problem for the exclusive-or predicate, which is known to be NP-hard [87]. VCSP(Γ_{xor}) is also equivalent to the NP-complete MAX-CUT problem [125]. For $|D| > 2$, the problem VCSP(Γ_{xor}) corresponds to the $|D|$-COLOURING problem, which is NP-complete. Therefore, Γ_{xor} is NP-hard (cf. Example 6).

Example 1.11 (Satisfiability) Consider a Boolean valued constraint language Γ. We can restrict Γ by allowing only cost functions with range $\{0, \infty\} \subseteq \overline{\mathbb{Q}}_{\geq 0}$. This way we obtain the standard SATISFIABILITY problem. The complexity of the VCSP for such restricted languages Γ has been completely characterised and six tractable classes have been identified [252].

Alternatively, if we restrict Γ by allowing only cost functions with range $\{0, 1\} \subseteq \overline{\mathbb{Q}}_{\geq 0}$, we obtain the MAX-SAT problem, in which the aim is to satisfy the maximum number of constraints. The complexity of this problem has been completely characterised and three tractable classes have been identified [87].

Example 1.12 (CSP) There has been a lot of research on the complexity of language-restricted CSPs. The first result in this line of research goes back to Schaefer, who obtained a complete classification of Boolean CSPs [252], as mentioned in Example 1.11. Other known complexity classification results include CSPs with a single binary symmetric relation (that is, graphs) [149] (cf. Example 1.19), CSPs with all unary relations (so-called *conservative* CSPs) [10, 45] (cf. Example 1.23), CSPs over a three-element domain [39], CSPs with a single binary relation without sources and sinks [14], and CSPs with a single binary relation that belongs to a class of oriented trees called special triads [13].

Remark 1.8 It is known that the CSP is equivalent to the HOMOMORPHISM problem between relational structures and many other problems [46, 60, 111, 150, 151, 161]. (In fact, the CSP is equivalent to the CSP with a single binary relation [150].) Restricting the class of source relational structures corresponds to structural restrictions, and restricting the class of target relational structures corresponds to language restrictions. Structure-restricted CSPs are also known as *uniform* CSPs, whereas language-restricted CSPs (with a fixed finite language) are known as *nonuniform* CSPs [184].

Feder and Vardi have shown that the nonuniform CSP is the largest subclass of NP that can exhibit a *dichotomy* [111]. In other words, nonuniform CSPs form the largest subclass of NP that might possibly contain only polynomial-time and NP-complete problems despite Ladner's Theorem, which shows existence of intermediate languages in NP provided $P \neq NP$ [199]. Feder and Vardi conjectured that the nonuniform CSP does indeed exhibit a dichotomy [111]. Bulatov et al. have given an algebraic characterisation of this *dichotomy conjecture* [42]. Various equivalent formulations of the dichotomy conjecture can be found in a beautiful survey paper by Hell and Nešetřil [151]; see also [226].

Feder and Vardi also showed that the nonuniform CSP is polynomial-time equivalent to a certain logic called MMSNP, that is, monotone monadic strict NP without inequality [111]. However, one of the reductions in [111] is a randomised polynomial-time reduction. More recently, Kun has proved a polynomial-time equivalence between the nonuniform CSP and MMSNP [195]. This equivalence has been refined by Kun and Nešetřil [196].

A variety of mathematical approaches to the nonuniform CSP have been suggested in the literature. The most advanced approaches use logic, combinatorics, universal algebra, and their combination [48, 54, 60, 137, 184].

The logic programming language DATALOG can be used to define CSPs of bounded width, which can be solved by local consistency algorithms [5, 94, 96, 162, 183]. Recent results have established the power of the local consistency technique for language-restricted CSPs [11, 201].[4]

In fact, all nonuniform CSPs that are known to be tractable can be solved via local consistency techniques, via the "few subpowers property" [16, 155] (which generalises Gaussian elimination and results in [40, 92, 111, 162]), or via a combination of the two.

More on connections between logic and *constraint satisfaction* can be found in [137, 151]. See also [60] for a survey on the complexity of constraint languages, and [88] for an overview of current research themes.

A result of Kun and Szegedy relates the dichotomy conjecture to continuous mathematics and techniques from PCPs (Probabilistically Checkable Proofs) [198].

Example 1.13 (Boolean Min/Max-CSP, Min/Max-Ones) Khanna et al. have obtained a complexity classification of Boolean MIN-CSP and MAX-CSP, and also Boolean MIN-ONES and MAX-ONES [179]; see also [87].

Example 1.14 (Boolean Min/Max-AW-CSP, Min/Max-AW-Ones) Consider a generalisation of the MIN-CSP and MAX-CSP frameworks that allows arbitrary weights, that is, both positive and negative weights. In terms of cost functions, this means that each cost function can take on values 0 and c for some fixed (positive or negative) c depending on the cost function. MIN-ONES and MAX-ONES are defined similarly.

Jonsson has generalised the results of Creignou et al. from Example 1.13 and has given a complexity classification of Boolean MIN-AW-CSP and MAX-AW-CSP, and also MIN-AW-ONES and MAX-AW-ONES [170].

Example 1.15 (Boolean VCSP) A complexity classification of Boolean VCSPs with arbitrary positive cost functions has been obtained by Cohen et al. [67]. We will describe this result in more detail in Chap. 6.

Example 1.16 (Non-Boolean Max-CSP) The first results on the MAX-CSP over arbitrary domains are due to Cohen et al. [59]. A complexity classification has been obtained for the MAX-CSP on three-element domains [171] and four-element domains [174].

Let Γ_{fix} be the language containing for every domain valued d the unary cost function defined by $u_d(d) = 1$ and $u_d(x) = 0$ otherwise (so-called *constant* or *fixed-value* constraints). A complexity classification of MAX-CSPs with fixed-value constraints, that is, languages including Γ_{fix}, with respect to approximability, has been obtained by Deineko et al. [100].

---

[4]The power of the local consistency technique has also been fully characterised for uniform CSPs [4].

See also [239] for results on the approximability and inapproximability of the
MAX-CSP. Barto and Kozik have recently linked robust satisfiability and semidefinite programming [12].

Example 1.17 (Max-AW-CSP) Jonsson and Krokhin have generalised results from
Example 1.14 from Boolean domains to arbitrary domains, and obtained a complexity classification of MAX-AW-CSPs, that is, MAX-CSPs with arbitrary weights
over arbitrary domains [172].

Example 1.18 (Max-Ones) Jonsson et al. have generalised the result from Example 1.13, and have obtained a complexity classification with respect to approximability of the MAX-ONES problem for maximal languages over domains of size up
to 4 and of the MAX-ONES problem with all permutation relations [173].

Notation 1.5 Given two graphs (undirected or directed) G and H, we denote by
$V(G)$ and $V(H)$ the set of vertices of G and H respectively. We denote by $E(G)$
and $E(H)$ the set of edges of G and H respectively. A mapping $f : V(G) \to V(H)$
is a *homomorphism* of G to H if f preserves edges, that is, $(u, v) \in E(G)$ implies
$(f(u), f(v)) \in E(H)$.

Example 1.19 (Graph homomorphism) The GRAPH HOMOMORPHISM problem
asks whether an input graph G admits a homomorphism to a fixed graph H. This
problem is also known as H-COLOURING [150].

  H-COLOURING is equivalent to VCSP(Γ_e), where Γ_e denotes the language containing a single binary symmetric relation representing the edges of H ("e" for
edge). A complexity classification of the H-COLOURING problem has been obtained by Hell and Nešetřil [149]: H-COLOURING is tractable if, and only if, H
contains a loop or H is bipartite; otherwise H-COLOURING is NP-complete. Bulatov has provided an algebraic proof of this result [43].

Example 1.20 (Graph list homomorphism) The GRAPH LIST HOMOMORPHISM
problem for H asks whether an input graph G with lists $L_u \subseteq V(H)$, $u \in V(G)$,
admits a homomorphism f to H such that $f(u) \in L_u$ for each $u \in V(G)$.

  Let Γ_{cons} consist of all unary relations ("cons" for conservative). Then
VCSP($\Gamma_e \cup \Gamma_{\text{cons}}$), where Γ_e is from Example 1.19, is equivalent to the GRAPH
LIST HOMOMORPHISM problem. A complexity classification of this problem is
due to Feder et al. [109].

Example 1.21 (Graph Min-Cost homomorphism) For two graphs G and H, consider nonnegative rational costs $c_v(u)$ for $u \in V(G)$ and $v \in V(H)$. The cost of an
homomorphism f of G to H is defined to be $\sum_{u \in V(G)} c_{f(u)}(u)$. For a fixed H, the
GRAPH MIN-COST HOMOMORPHISM problem asks to find a homomorphism of G
to H with minimum cost.

  GRAPH MIN-COST HOMOMORPHISM is equivalent to VCSP($\Gamma_e \cup \Gamma_{\text{scons}}$),
where Γ_e is from Example 1.19, and Γ_{scons} consists of all unary cost functions

("scons" for soft conservative). A complexity classification of this problem is due
to Gutin et al. [143].

Remark 1.9 Structurally-restricted variants of the GRAPH HOMOMORPHISM problems from Examples 1.19, 1.20 and 1.21 have also been studied; see [107, 139].

Example 1.22 (Digraph homomorphism) The DIGRAPH HOMOMORPHISM problem is an analogue of GRAPH HOMOMORPHISM from Example 1.19 for directed
graphs. As mentioned in Remark 1.8, this problem captures the complexity of
language-restricted CSPs.

Let Γ_a denote the language containing a single binary relation ("a" for arc). For
any fixed Γ_a, VCSP(Γ_a) is polynomial-time equivalent to the DIGRAPH HOMO-
MORPHISM problem for the graph whose edge relation is given by the binary relation from Γ_a [150]. A complexity classification of the DIGRAPH HOMOMORPHISM
problem for semicomplete digraphs has been obtained in [9].

Example 1.23 (Digraph list homomorphism) The DIGRAPH LIST HOMOMOR-
PHISM problem is an analogue of GRAPH LIST HOMOMORPHISM from Example 1.20 for directed graphs.

VCSP($\Gamma_a \cup \Gamma_{cons}$), where Γ_a is from Example 1.22, and Γ_{cons} is from Example 1.20, is equivalent to the DIGRAPH LIST HOMOMORPHISM problem. Bulatov
has obtained a complexity classification of this problem [45].

Example 1.24 (Digraph Min-Cost homomorphism) The DIGRAPH MIN-COST
HOMOMORPHISM problem is an analogue of GRAPH MIN-COST HOMOMOR-
PHISM from Example 1.21 for directed graphs.

VCSP($\Gamma_a \cup \Gamma_{sconst}$), where Γ_a is from Example 1.22 and Γ_{sconst} is from Example 1.21, is equivalent to the DIGRAPH MIN-COST HOMOMORPHISM problem.
This problem is also equivalent to the LEVEL OF REPAIR ANALYSIS problem [145].
In a long series of paper, Gutin et al. have classified this problem for various special classes of digraphs [144]. All these partial results have been generalised by
Takhanov, who has provided a complete classification in [261], with generalisations
in [262].

Example 1.25 (Conservative VCSP) A complexity classification of conservative
VCSPs has been obtained by Kolmogorov and Živný [187, 188]; these are VCSPs
containing all unary {0, 1}-valued cost functions. (In fact, their result also provides
a complexity classification of fixed-valued languages; cf. Example 1.16.) Their result generalises results from Example 1.16 (with respect to exact solvability), (some
results from) Examples 1.15, 1.21, and 1.24. We will discuss their result and techniques in more detail in Chap. 7.

Example 1.26 (Max-Sol) The MAXIMUM SOLUTION problem is equivalent to
the VCSP over the language consisting of all relations and unary cost functions with the following cost functions: $\mu(d) = wd$ for any domain value d and

some fixed $w \in \mathbb{N}$ [175]. Jonsson et al. have studied the MAX-SOL problem over graphs [176, 177], which is a restriction of GRAPH MIN-COST HOMOMORPHISM from Example 1.19, but a generalisation of both GRAPH LIST HOMOMORPHISM from Example 1.20 and MAX-ONES from Example 1.18.

1.3.3 Hybrid Restrictions

Surprisingly, until very recently not much research has been done on the combination of the two mentioned restrictions (that is, language and structural restrictions), which would result in finding *hybrid* reasons for tractability of VCSPs. Some results have appeared on hybrid tractability of CSPs [62, 64, 74, 76]. We will discuss hybrid tractability of VCSPs in more detail in Chap. 9.

1.3.4 Related Work

Let us mention some related areas that have been considered in the literature. The first three are extensions of CSPs dealing with decision problems.

Firstly, CSPs over infinite domains have been studied. Complexity classifications have been obtained for subsets of Allen's interval algebra [190, 223], equality constraint languages [25], and temporal CSPs [26]. We refer the reader to a survey for more information [22].

Secondly, quantified CSPs have been studied. Creignou et al. have studied quantified CSPs over Boolean domains [87]. More work on quantified CSPs was initiated by [32] and Chen's thesis [53]; see also [55]. A complexity classification has been obtained for quantified equality constraint languages [23] and relatively quantified CSPs [24, 110] (that is, quantified CSPs with all unary constraints; in the case of non-quantified CSPs, these are known as conservative CSPs). We refer the reader to [54] for more information.

Thirdly, language-restricted CSPs correspond to evaluating the primitive positive fragment $(\exists, \wedge)$ of first-order (FO) logic over a fixed finite structure. Moreover, language-restricted quantified CSPs correspond to evaluating the positive Horn fragment $(\exists, \forall, \wedge)$ of FO over a fixed finite structure. Martin has shown that, apart from CSPs and quantified CSPs, there is only one more interesting, with respect to computational complexity, fragment of FO, namely the positive first-order logic without equality $(\exists, \forall, \wedge, \vee)$ [211]. A complete complexity classification has been obtained for domains of size at most 3 [209] and then for all finite domains [208].

Finally, in the COUNTING CONSTRAINT SATISFACTION problem, #CSP, the goal is to find the number of solutions. The general framework is similar to the VCSP; instead of minimising the sum of cost functions (over all possible assignments of values to variables), the objective is to compute the sum (again, over all possible assignments of values to variables) of the product of all $\{0, 1\}$-valued cost functions. Formally, for a #CSP instance $\mathcal{P} = \langle V, D, \mathcal{C} \rangle$, the goal is to compute

$$\mathrm{Eval}(\mathcal{P}) \stackrel{\mathrm{def}}{=} \sum_{s \in \mathcal{A}} \prod_{\langle \mathbf{x}, \phi \rangle \in \mathcal{C}} \phi(\mathbf{x}).$$

Recall from Definition 1.1 that $\mathcal{A}$ denotes the set of all assignments of values to the variables.

From the many results in this area, let us just mention the complexity classification of #CSPs [44, 103].

The COUNTING CONSTRAINT SATISFACTION problem can be further generalised from $\{0, 1\}$-valued cost functions to arbitrary cost functions. This is known as the PARTITION FUNCTION problem, the WEIGHTED #CSP problem or just #CSP. Recall that the goal is to compute the sum (over all possible assignments of values to variables) of the product of all cost functions in a given instance. Note that in this case the resulting number does not correspond to the number of solutions anymore as the concept of number of solutions does not make any sense.

From a long series of results, let us mention the latest result that, apart from generalising all previous results, provides a dichotomy theorem for all partition functions with complex weights [50].

1.4 Expressibility

In any VCSP instance, the variables listed in the scope of each valued constraint are *explicitly* constrained, in the sense that each possible combination of values for those variables is associated with a given cost. Moreover, if we choose *any* subset of all variables, then their values are constrained *implicitly* in the same way, due to the combined effect of the constraints.

Definition 1.4 (Expressibility) For any VCSP instance $\mathcal{P} = \langle V, D, \mathcal{C} \rangle$, and any m-tuple $\mathbf{v} = \langle v_1, \ldots, v_m \rangle$ of variables of $\mathcal{P}$, the *projection* of $\mathcal{P}$ onto $\mathbf{v}$, denoted by $\pi_{\mathbf{v}}(\mathcal{P})$, is the m-ary cost function defined by

$$\pi_{\mathbf{v}}(\mathcal{P})(\mathbf{x}) \stackrel{\mathrm{def}}{=} \min_{s \in \mathcal{A}} \{ \mathrm{Cost}_{\mathcal{P}}(s) \mid s(\mathbf{v}) = \mathbf{x} \},$$

where $\mathcal{A}$ denotes the set of all assignments (cf. Definition 1.1). We say that a cost function ϕ is *expressible* over a valued constraint language Γ if there exists an instance $\mathcal{P}$ of VCSP(Γ) and a tuple $\mathbf{v}$ of variables of $\mathcal{P}$ such that $\pi_{\mathbf{v}}(\mathcal{P}) = \phi$. The variables of $\mathcal{P}$ not from $\mathbf{v}$ are called *extra* (or *hidden*) variables. We call the pair $\langle \mathcal{P}, \mathbf{v} \rangle$ a *gadget* for expressing ϕ over Γ.

Note that the tuple of variables $\mathbf{v}$ in a gadget may contain repeated entries, the sum over an empty set is 0, and the minimum over an empty set is ∞.

Expressibility allows a particular form of problem reduction: if a constraint can be expressed in a given constraint language, then it can be added to the language without changing the computational complexity of the associated class of problems (cf. Theorem 1.1 below). To indicate its central role we note that the same basic idea of expressibility has been studied under many different names in different fields: implementation [87], pp-definability [54], existential inverse satisfiability [89], structure identification [99], and join and projection operations in relational databases [266]. In the context of the CSP, both upper bounds [164, 166] and lower bounds [270] on the complexity of deciding expressibility have been studied.

Example 1.27 Let $\mathcal{P}$ be the VCSP instance with a single variable v and no constraints, and let $\mathbf{v} = \langle v, v \rangle$. Then, by Definition 1.4,

$$\pi_{\mathbf{v}}(\mathcal{P})(x, y) = \begin{cases} 0 & \text{if } x = y, \\ \infty & \text{otherwise.} \end{cases}$$

Hence for any valued constraint language Γ, over any set D, the binary *equality* relation is expressible over Γ.

Example 1.28 Let D be a finite set of size d. Consider a valued constraint language $\Gamma = \{\neq_d\}$ over D that consists of a binary disequality relation, $\neq_d$, given by

$$\neq_d \overset{\text{def}}{=} \{\langle a, b \rangle \in D^2 \mid a \neq b\}.$$

Note that $\neq_d$ is a "hard variant" of ϕ_{xor} from Example 1.10. Consider an instance $\mathcal{P} = \{V, D, \mathcal{C}\}$ of $\mathsf{VCSP}(\Gamma)$, where $V = \{x_1, \ldots, x_{n+1}\}$, $n = d$, and

$$\mathcal{C} = \{\langle \langle x_i, x_j \rangle, \neq_d \rangle \mid i \neq j \in \{1, \ldots, n\}\} \cup \{\langle \langle x_i, x_{n+1} \rangle, \neq_d \rangle \mid i \in \{2, \ldots, n\}\}.$$

In order to satisfy all constraints from $\mathcal{C}$, variables $x_1, \ldots, x_n$ have to be assigned different values. Moreover, the value of the variable x_{n+1} has to be different from the values of the variables $x_2, \ldots, x_n$. Hence, the only remaining value that can be assigned to the variable x_{n+1} is the value which is assigned to the variable x_1. Therefore, every solution s to $\mathcal{P}$ with minimum total cost (in this case 0) satisfies $s(x_1) = s(x_{n+1})$. Therefore, $\langle \mathcal{P}, \{x_1, x_{n+1}\} \rangle$ is a gadget for the equality relation $=_d$, given by

$$=_d \overset{\text{def}}{=} \{\langle a, b \rangle \in D^2 \mid a = b\}.$$

In other words, the equality relation can be expressed using the disequality relation. An example of this construction for $|D| = 3$ is shown in Fig. 1.2.

Example 1.29 Consider the VCSP instance $\mathcal{P}$ from Example 1.7. The projection of $\mathcal{P}$ onto $\langle x_2, x_4 \rangle$, denoted by $\pi(\mathcal{P})_{\langle x_2, x_4 \rangle}$, is a binary cost function defined by minimising over the remaining variables. The following table, which enumerates all assignments s in which x_2 and x_4 are both assigned 0, together with the cost of these assignments, shows that $\pi(\mathcal{P})_{\langle x_2, x_4 \rangle}(0, 0) = 21$.

Fig. 1.2 The gadget
expressing $=_3$ over $\{\neq_3\}$ for
$|D| = 3$ (Example 1.28)

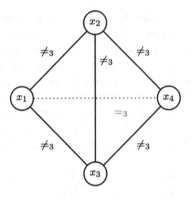

x_2	x_4	x_1	x_3	x_5	$Cost_{\mathcal{P}}(s)$
0	0	0	0	0	23
0	0	0	0	1	22
0	0	0	1	0	24
0	0	**0**	**1**	**1**	**21**
0	0	1	0	0	25
0	0	1	0	1	24
0	0	1	1	0	26
0	0	1	1	1	23

Similarly, it is straightforward to check that

$$\pi(\mathcal{P})_{\langle x_2, x_4 \rangle}(x, y) = \begin{cases} 21 & \text{if } x = 0 \text{ and } y = 0, \\ 16 & \text{if } x = 0 \text{ and } y = 1, \\ 24 & \text{if } x = 1 \text{ and } y = 0, \\ 19 & \text{if } x = 1 \text{ and } y = 1. \end{cases}$$

Hence this cost function can be expressed over the valued constraint language Γ
defined in Example 1.7.

Example 1.30 Consider a ternary finite-valued cost function ϕ over $D = \{0, 1, 2\}$
defined by $\phi = (\#0)^2$, that is, the square of the number of 0s in the input. We will
construct a gadget for expressing ϕ using only binary crisp cost functions and finite-
valued unary cost functions.

Define three binary crisp cost functions as follows:

$$\phi_0(x, y) \stackrel{\text{def}}{=} \begin{cases} \infty & \text{if } x = 0 \text{ and } y = 1, \\ \infty & \text{if } x = 0 \text{ and } y = 2, \\ 0 & \text{otherwise.} \end{cases}$$

$$\phi_1(x, y) \stackrel{\text{def}}{=} \begin{cases} \infty & \text{if } x = 0 \text{ and } y = 1, \\ 0 & \text{otherwise,} \end{cases}$$

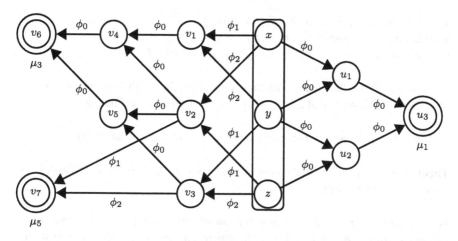

Fig. 1.3 The gadget expressing $\phi = (\#0)^2$ (Example 1.30)

and

$$\phi_2(x, y) \stackrel{\text{def}}{=} \begin{cases} \infty & \text{if } x = 0 \text{ and } y = 2, \\ 0 & \text{otherwise.} \end{cases}$$

For $c \in \{1, 3, 5\}$, let μ_c be a unary finite-valued cost function defined by

$$\mu_c(x) \stackrel{\text{def}}{=} \begin{cases} c & \text{if } x = 0, \\ 0 & \text{otherwise.} \end{cases}$$

Let $\mathcal{P} = \langle V, D, \mathcal{C} \rangle$ where $V = \{x, y, z, u_1, u_2, u_3, v_1, v_2, v_3, v_4, v_5, v_6, v_7\}$ and the set of constraints $\mathcal{C}$ is as shown in Fig. 1.3.

We claim that $\langle \mathcal{P}, \langle x, y, z \rangle \rangle$ is a gadget for expressing ϕ.

If all x, y, and z are nonzero, then there is an assignment of the other variables with values 1 and 2 such that the total cost is 0.

If any of x, y, z is 0, then in any minimum-cost assignment either u_1 or u_2 is assigned 0, and for the same reason u_3 is assigned 0.

If at least two of x, y, z are 0, then in any minimum-cost assignment at least one of the variables v_1, v_2, v_3 is assigned 0, and consequently, at least one of v_4, v_5 is assigned 0, and hence v_6 is assigned 0.

If all x, y, and z are 0, then both v_2 and v_3 are assigned 0, and consequently, v_7 is assigned 0.

Note that a similar gadget can be constructed for larger domains.

In view of Example 1.27, we will assume that every valued constraint language contains the binary equality relation.

Definition 1.5 (Expressive power) For any valued constraint language Γ, we define the *expressive power* of Γ, denoted by $\langle \Gamma \rangle$, to be the closure of Γ including the binary equality relation and closed under:

- scaling: $\phi \in \langle \Gamma \rangle, \alpha \in \mathbb{Q}_{\geq 0}, \beta \in \mathbb{Q} \Rightarrow \alpha\phi + \beta \in \langle \Gamma \rangle$
- addition of cost function: $\phi, \psi \in \langle \Gamma \rangle \Rightarrow \phi + \psi \in \langle \Gamma \rangle$
- expressibility: $\phi \in \langle \Gamma \rangle, \psi(\mathbf{x}) = \min_{\mathbf{y}} \phi(\mathbf{x}, \mathbf{y}) \Rightarrow \psi \in \langle \Gamma \rangle$

Remark 1.10 Note that scaling implicitly includes the feasibility operator as we have defined $0\infty = \infty$; that is, for every cost function ϕ, $0\phi = \mathsf{Feas}(\phi)$.

The following theorem shows that expressibility preserves tractability.

Theorem 1.1 *For any valued constraint language* Γ, $\mathsf{VCSP}(\Gamma)$ *and* $\mathsf{VCSP}(\langle \Gamma \rangle)$ *are log-space equivalent.*

Remark 1.11 Note that the original version of this result from [67] shows only polynomial-time equivalence. However, this result can be extended due to Reingold's result that (s, t)-CONNECTIVITY in undirected graphs can be solved in logarithmic space [244].

Example 1.31 By Theorem 1.1 and Example 1.10, in order to show that Γ is an NP-hard valued constraint language it is sufficient to show that ϕ_{xor} is expressible over Γ.

1.5 Algebraic Properties

In this section, we will describe algebraic techniques that have been developed for determining the expressive power of valued constraint languages. To make use of these techniques, we first need to define some key terms.

Any operation on a set D can be extended to tuples over the set D in a standard way, as follows. For any function $f : D^k \to D$, and any collection of tuples $t_1, \ldots, t_k \in D^m$, define $f(t_1, \ldots, t_k) \in D^m$ to be the tuple $\langle f(t_1[1], \ldots, t_k[1]), \ldots, f(t_1[m], \ldots, t_k[m]) \rangle$. For instance, if $f : D^2 \to D$, then $f(t_1, t_2) = \langle f(t_1[1], t_2[1]), \ldots, f(t_1[k], t_2[k]) \rangle$ for any $t_1, t_2 \in D^k$.

Definition 1.6 (Polymorphism [101]) Let R be an m-ary relation over a finite set D and let f be a k-ary operation on D. Then f is a *polymorphism* of R if $f(t_1, \ldots, t_k) \in R$ for all choices of $t_1, \ldots, t_k \in R$.

Notation 1.6 We denote by $e_i^{(k)} : D^k \to D$ the *projection* operation that returns its ith argument, that is, $e_i^{(k)}(x_1, \ldots, x_n) = x_i$.

Observation 1.1 It follows readily from the definitions that any projection $e_i^{(k)}$ is a polymorphism of all relations.

Notation 1.7 We will say that f is a polymorphism of a crisp constraint language Γ if f is a polymorphism of every relation in Γ. We will say that f is a polymorphism of a (not necessarily crisp) language Γ if f is a polymorphism of $\mathsf{Feas}(\Gamma)$, where $\mathsf{Feas}(\Gamma) = \{\mathsf{Feas}(\phi) \mid \phi \in \Gamma\}$.[5]

Notation 1.8 The set of all polymorphisms of Γ will be denoted by $\mathsf{Pol}(\Gamma)$.

It follows from the results of Geiger and Bodnarčuk et al. that the expressive power of a crisp constraint language is fully characterised by its polymorphisms:

Theorem 1.2 ([27, 127, 163]) *For any crisp constraint language Γ over a finite set,*

$$R \in \langle \Gamma \rangle \quad \Leftrightarrow \quad \mathsf{Pol}(\Gamma) \subseteq \mathsf{Pol}(\{R\}).$$

Hence, a relation R is expressible over a crisp constraint language Γ if, and only if, it has all the polymorphisms of Γ. See [165] for more on the connection between crisp constraint languages on the one hand, and universal algebra on the other hand.

Remark 1.12 It is known that crisp constraint languages that have only trivial polymorphisms are NP-hard [42], where trivial means projections and semi-projections. (Note that this is a powerful technique for showing NP-hardness results, as one does not need to build a reduction from some known NP-hard problem.) The famous *dichotomy conjecture*, mentioned in Remark 1.8, can be equivalently stated as follows: if a crisp constraint language Γ has a nontrivial polymorphism, then $\mathsf{CSP}(\Gamma)$ is polynomial-time solvable. In fact, there is only one remaining type of polymorphism for which we currently do not know whether it leads to polynomial or NP-hard problems, and that is *Taylor* polymorphism [151]. Establishing the complexity of crisp constraint languages with a Taylor polymorphism would resolve the dichotomy conjecture. Maróti and McKenzie have shown that Taylor polymorphisms are equivalent, with respect to the question of tractability, to so-called *weak near-unanimity* polymorphisms [210].

In order to capture the expressive power of valued languages, we need a generalisation of polymorphisms. We start with binary multimorphisms [67].

Definition 1.7 (Binary multimorphism) Let $\langle f, g \rangle$ be a pair of operations $f, g : D^2 \rightarrow D$. We say that an m-ary cost function $\phi : D^m \rightarrow \overline{\mathbb{Q}}_{\geq 0}$ admits $\langle f, g \rangle$ as a *multimorphism* if

$$\phi\big(f(x_1, y_1), \ldots, f(x_m, y_m)\big) + \phi\big(g(x_1, y_1), \ldots, g(x_m, y_m)\big)$$

$$\leq \phi(x_1, \ldots, x_m) + \phi(y_1, \ldots, y_m) \tag{1.1}$$

for all $x_1, y_1, \ldots, x_m, y_m \in D$.

---

[5]Polymorphisms of valued constraint languages are also known as *feasibility polymorphisms* [65].

If a cost function ϕ admits $\langle f, g \rangle$ as a multimorphism, we also say that ϕ is *improved* by $\langle f, g \rangle$. We say that a language Γ admits $\langle f, g \rangle$ as a multimorphism (or equivalently, that Γ is improved by $\langle f, g \rangle$) if every cost function ϕ from Γ admits $\langle f, g \rangle$ as a multimorphism.

Remark 1.13 Using standard vector notation, an m-ary cost function ϕ admits $\langle f, g \rangle$ as a multimorphism if

$$\phi\big(f(\mathbf{x}, \mathbf{y})\big) + \phi\big(g(\mathbf{x}, \mathbf{y})\big) \leq \phi(\mathbf{x}) + \phi(\mathbf{y})$$

for all $\mathbf{x}, \mathbf{y} \in D^m$, where both f and g are applied coordinatewise.

Binary multimorphisms are important as they are a natural generalisation of submodularity, which is defined in Sect. 1.6, and also define many tractable languages, as discussed in Chap. 6.

It is natural to consider the case of more than two closure functions.

Definition 1.8 (General multimorphism) Let $\mathbf{g} = \langle g_1, \ldots, g_k \rangle$ be a k-tuple of k-ary operations $g_i : D^k \to D$, $1 \leq i \leq k$. An m-ary cost function $\phi : D^m \to \overline{\mathbb{Q}}_{\geq 0}$ admits $\mathbf{g}$ as a multimorphism if

$$\sum_{i=1}^{k} \phi\big(g_i(\mathbf{x_1}, \ldots, \mathbf{x_k})\big) \leq \sum_{i=1}^{k} \phi(\mathbf{x_i}) \qquad (1.2)$$

for all $\mathbf{x}_i \in D^m$, $1 \leq i \leq k$, where all functions g_i, $1 \leq i \leq k$, are applied coordinatewise.

Definition 1.8 is illustrated in Fig. 1.4, which should be read from left to right. Starting with the m-tuples $\mathbf{x}_1, \ldots, \mathbf{x}_k$, we first apply functions $g_1, \ldots, g_k$ on these tuples coordinatewise, thus obtaining the m-tuples $\mathbf{x}'_1, \ldots, \mathbf{x}'_k$. Inequality 1.2 amounts to comparing the sum of ϕ applied to tuples $\mathbf{x}_1, \ldots, \mathbf{x}_k$ and ϕ applied to tuples $\mathbf{x}'_1, \ldots, \mathbf{x}'_k$. (We call such a table a *tableau*.)

Remark 1.14 A cost function $\phi : D^m \to \overline{\mathbb{Q}}_{\geq 0}$ has a multimorphism $\mathbf{g}$ if ϕ satisfies a set of linear inequalities. Hence, from the geometrical point of view, for each possible fixed arity m, $\mathbf{g}$ corresponds to a hyperplane in a space of dimension $|D|^m$ [256].

However, to fully capture the expressive power of valued constraint languages it is necessary to consider more general algebraic properties. We start with fractional operations [65].

Definition 1.9 (Fractional operation) A k-ary *fractional operation* $\mathcal{F}$ on a set D is a set of the form $\{\langle r_1, f_1 \rangle, \ldots, \langle r_n, f_n \rangle\}$ where each r_i is a positive rational number such that $\sum_{i=1}^{n} r_i = k$ and each f_i is a distinct function from D^k to D.

$$
\begin{array}{l}
\mathbf{x_1} \\
\mathbf{x_2} \\
\vdots \\
\mathbf{x_k}
\end{array}
\qquad
\begin{array}{cccc}
\mathbf{x_1}[1] & \mathbf{x_1}[2] & \cdots & \mathbf{x_1}[m] \\
\mathbf{x_2}[1] & \mathbf{x_2}[2] & \cdots & \mathbf{x_2}[m] \\
& \vdots & & \\
\mathbf{x_k}[1] & \mathbf{x_k}[2] & \cdots & \mathbf{x_k}[m]
\end{array}
\quad \overset{\phi}{\longrightarrow} \quad
\left. \begin{array}{l}
\phi(\mathbf{x_1}) \\
\phi(\mathbf{x_2}) \\
\vdots \\
\phi(\mathbf{x_k})
\end{array} \right\}
\sum_{i=1}^{k} \phi(\mathbf{x_i})
$$

IV

$$
\begin{array}{l}
\mathbf{x_1'} = g_1(\mathbf{x_1},\ldots,\mathbf{x_k}) \\
\mathbf{x_2'} = g_2(\mathbf{x_1},\ldots,\mathbf{x_k}) \\
\vdots \\
\mathbf{x_k'} = g_k(\mathbf{x_1},\ldots,\mathbf{x_k})
\end{array}
\qquad
\begin{array}{cccc}
\mathbf{x_1'}[1] & \mathbf{x_1'}[2] & \cdots & \mathbf{x_1'}[m] \\
\mathbf{x_2'}[1] & \mathbf{x_2'}[2] & \cdots & \mathbf{x_2'}[m] \\
& \vdots & & \\
\mathbf{x_k'}[1] & \mathbf{x_k'}[2] & \cdots & \mathbf{x_k'}[m]
\end{array}
\quad \overset{\phi}{\longrightarrow} \quad
\left. \begin{array}{l}
\phi(\mathbf{x_1'}) \\
\phi(\mathbf{x_2'}) \\
\vdots \\
\phi(\mathbf{x_k'})
\end{array} \right\}
\sum_{i=1}^{k} \phi(\mathbf{x_i'})
$$

Fig. 1.4 Definition of a multimorphism $\mathbf{g} = \langle g_1, \ldots, g_k \rangle$

Now we generalise multimorphisms to operations with weights [65].

Definition 1.10 (Fractional polymorphism) For any m-ary cost function ϕ, we say that a k-ary fractional operation $\mathcal{F} = \{\langle f_1, f_1 \rangle, \ldots, \langle r_n, f_n \rangle\}$ is a k-ary *fractional polymorphism* of ϕ if

$$
\sum_{i=1}^{n} r_i \phi\big(f_i(\mathbf{x_1}, \ldots, \mathbf{x_k}) \big) \leq \sum_{i=1}^{k} \phi(\mathbf{x_i}) \tag{1.3}
$$

for all $\mathbf{x_i} \in D^m$, $1 \leq i \leq k$.

Definition 1.10 is illustrated in Fig. 1.5.

Remark 1.15 A k-ary fractional polymorphism whose weights are all natural numbers is a multimorphism.

Remark 1.16 An equivalent definition of a fractional polymorphism is given in [65]: all weights have to be natural numbers; however, there is no restriction on the sum of all weights. Such a fractional polymorphism is a multimorphism if the sum of all weights is equal to k.

Remark 1.17 An equivalent definition[6] of fractional operations and fractional polymorphisms is the following. A k-ary fractional operation defined on D is a probability distribution given on $\mathbf{O}_D^{(k)}$, the set of all k-ary operations on D. A k-ary fractional operation $\mathcal{F}$ is a fractional polymorphism of an m-ary cost function ϕ if, for all $\mathbf{x_1}, \ldots, \mathbf{x_k} \in D^m$,

$$
\mathbb{E}_{f \sim \mathcal{F}}\big(\phi\big(f(\mathbf{x_1}, \ldots, \mathbf{x_k}) \big) \big) \leq \operatorname{avg}\{\phi(\mathbf{x_1}), \ldots, \phi(\mathbf{x_k})\},
$$

---

[6]For a k-ary fractional operation, all weights r_i are divided by k.

$$
\begin{array}{cccccc}
\mathbf{x}_1 & \mathbf{x}_1[1] & \mathbf{x}_1[2] & \cdots & \mathbf{x}_1[m] & \phi(\mathbf{x}_1) \\
\mathbf{x}_2 & \mathbf{x}_2[1] & \mathbf{x}_2[2] & \cdots & \mathbf{x}_2[m] & \phi(\mathbf{x}_2) \\
\vdots & & \vdots & & & \vdots \\
\mathbf{x}_k & \mathbf{x}_k[1] & \mathbf{x}_k[2] & \cdots & \mathbf{x}_k[m] & \phi(\mathbf{x}_k)
\end{array}
\xrightarrow{\phi}
\left.\begin{array}{c}\phi(\mathbf{x}_1)\\\phi(\mathbf{x}_2)\\\vdots\\\phi(\mathbf{x}_k)\end{array}\right\}\sum_{i=1}^{k}\phi(\mathbf{x}_i)
$$

IV

$$
\begin{array}{cccccc}
\mathbf{x}'_1 = f_1(\mathbf{x}_1,\ldots,\mathbf{x}_k) & \mathbf{x}'_1[1] & \mathbf{x}'_1[2] & \cdots & \mathbf{x}'_1[m] & \phi(\mathbf{x}'_1) \\
\mathbf{x}'_2 = f_2(\mathbf{x}_1,\ldots,\mathbf{x}_k) & \mathbf{x}'_2[1] & \mathbf{x}'_2[2] & \cdots & \mathbf{x}'_2[m] & \phi(\mathbf{x}'_2) \\
\vdots & & \vdots & & & \vdots \\
\mathbf{x}'_n = f_k(\mathbf{x}_1,\ldots,\mathbf{x}_k) & \mathbf{x}'_n[1] & \mathbf{x}'_k[2] & \cdots & \mathbf{x}'_k[m] & \phi(\mathbf{x}'_n)
\end{array}
\xrightarrow{\phi}
\left.\begin{array}{c}\phi(\mathbf{x}'_1)\\\phi(\mathbf{x}'_2)\\\vdots\\\phi(\mathbf{x}'_n)\end{array}\right\}\sum_{i=1}^{k}r_i\phi(\mathbf{x}'_i)
$$

Fig. 1.5 Definition of a fractional polymorphism $\mathcal{F} = \{\langle r_1, f_1\rangle, \ldots, \langle r_n, f_n\rangle\}$

which is,

$$
\sum_{f\in O_D^{(k)}} \mathrm{Pr}_{\mathcal{F}}[f] \cdot \phi\big(f(\mathbf{x}_1,\ldots,\mathbf{x}_k)\big) \leq \frac{1}{k}\big(\phi(\mathbf{x}_1)+\cdots+\phi(\mathbf{x}_k)\big).
$$

A k-ary fractional polymorphism $\mathcal{F}$ is a multimorphism if the probability of each operation in $\mathcal{F}$ is of the form ℓ/k for some integer ℓ.

Notation 1.9 For any set of cost functions Γ, we denote by $\mathsf{fPol}(\Gamma)$ the set of all $\mathcal{F}$ such that $\mathcal{F}$ is a fractional polymorphism of every cost function in Γ. Similarly, $\mathsf{Mul}(\Gamma)$ denotes the set of multimorphisms of all cost functions from Γ.

Notation 1.10 For any set of fractional operations Ω defined over D, we denote by $\mathsf{Imp}(\Omega)$ the set Γ of cost functions defined over D such that any fractional operation $\mathcal{F} \in \Omega$ is a fractional polymorphism of all cost functions from Γ. (Imp is for "improved".) We will write $\mathsf{Imp}(\mathcal{F})$ for $\mathsf{Imp}(\{\mathcal{F}\})$ if $\mathcal{F}$ is a single fractional operation.

Observation 1.2 It is a simple consequence of the definitions that if $\mathcal{F} = \{\langle r_1, f_1\rangle, \ldots, \langle r_n, f_n\rangle\}$ is a fractional polymorphism of ϕ, then $\{f_i\}_{1\leq i\leq n}$ are polymorphisms of ϕ. On the other hand, if $\{f_i\}_{1\leq i\leq k}$ are polymorphisms of ϕ, then $\langle f_1, \ldots, f_k\rangle$ is not necessarily a multimorphism, and therefore not necessarily a fractional polymorphism, of ϕ. However, in the case of crisp cost functions the relationship is tighter. If $\{f_i\}_{1\leq i\leq n}$ are polymorphisms of a crisp cost function ϕ, then any fractional operation $\{\langle r_1, f_1\rangle, \ldots, \langle r_n, f_n\rangle\}$ is a fractional polymorphism of ϕ.

Observation 1.3 ([67]) If Γ is a tractable valued constraint language, then the set of relations $\{\mathsf{Feas}(\phi) \mid \phi \in \Gamma\}$ must be a tractable crisp constraint language.

Example 1.32 A *multi-projection* is a mapping from D^k to D^k that only permutes the set of its arguments. It follows from Definition 1.8 that every multi-projection is a multimorphism of all cost functions.

It has been shown in [65] that the polymorphisms and fractional polymorphisms of a valued constraint language effectively determine its expressive power. One consequence of this result is the following theorem:

Theorem 1.3 ([65]) *If Γ is a valued constraint language containing a constant function and closed under scaling, then*

$$\phi \in \langle \Gamma \rangle \quad \Leftrightarrow \quad \mathsf{Pol}(\Gamma) \subseteq \mathsf{Pol}(\{\phi\}) \wedge \mathsf{fPol}(\Gamma) \subseteq \mathsf{fPol}(\{\phi\}).$$

Corollary 1.1 *For all suitable valued constraint languages Γ, a cost function ϕ is expressible over Γ if, and only if, it has all the polymorphisms and fractional polymorphisms of Γ.*

We finish this section with a simple example of the algebraic technique. We have shown in Example 1.28 that the disequality relation can express the equality relation. We now investigate the converse.

Example 1.33 Consider a constraint language $\Gamma = \{=_d\}$ over D, $|D| = d$, which consists of the binary equality relation $=_d$ from Example 1.28. Consider k arbitrary 2-tuples $t_1, \ldots, t_k$ over D and an arbitrary function $f : D^k \to D$. If $t_i \in \{=_d\}$ for every $i = 1, \ldots, k$, then $f(t_1[1], \ldots, t_k[1]) = f(t_1[2], \ldots, t_k[2])$ and therefore $f(t_1, \ldots, t_k) \in \{=_d\}$. It follows that every function is a polymorphism of $=_d$. Obviously, not every function is a polymorphism of $\neq_d$: a simple counterexample is a constant function. We have shown that $\mathsf{Pol}(\{=_d\}) \not\subseteq \mathsf{Pol}(\{\neq_d\})$ and therefore $\neq_d$ is not expressible over $\{=_d\}$ by Theorem 1.2. This is almost obvious, but this simple example illustrates the use of the algebraic approach.

1.6 Submodularity

For any finite set V, a function $f : 2^V \to \mathbb{Q}_{\geq 0}$ defined on subsets[7] of V is called a *set function*.

Definition 1.11 (Submodularity on sets) A set function $f : 2^V \to \mathbb{Q}_{\geq 0}$ is called *submodular* if for all $S, T \subseteq V$,

$$f(S \cap T) + f(S \cup T) \leq f(S) + f(T).$$

Remark 1.18 An equivalent definition of submodularity is the property of *decreasing marginal values:* for any $A \subseteq B \subseteq V$ and $x \in V$,

$$f(B \cup \{x\}) - f(B) \leq f(A \cup \{x\}) - f(A).$$

---

[7]Think of 2^V as $\{0, 1\}$-vectors of length $|V|$.

This can be deduced from the first definition by substituting $S = A \cup \{x\}$ and $T = B$; the reverse implication also holds [258] and can be shown by induction on $|A \cup B| - |A \cap B|$.

Submodular functions are a key concept in operational research and combinatorial optimisation [121, 158, 189, 222, 225, 258, 265]. Examples include cuts in graphs [130, 238], matroid rank functions [106], set covering problems [112], and entropy functions. Submodular functions are often considered to be a discrete analogue of convex functions [205].

Both minimising and maximising submodular functions, possibly under some additional conditions, have been considered extensively in the literature. Most scenarios use the so-called *oracle value model*: for any set S, an algorithm can query an oracle to find the value of $f(S)$.

Submodular function maximisation is easily shown to be NP-hard [258] since it generalises many standard NP-hard problems such as the MAX-CUT problem; see also [114]. In contrast, the SUBMODULAR FUNCTION MINIMISATION problem, SFM, which consists in *minimising* a submodular function, can be solved efficiently with only polynomially many oracle calls.

The first polynomial algorithm for the SFM problem is due to Grötschel, Lovász, and Schrijver [141]. A strongly[8] polynomial algorithm has been described in [142]. These algorithms employ the ellipsoid method.

Based on the work of Cunningham [90, 91], several combinatorial algorithms have been obtained in the last decade [116, 156, 157, 159, 160, 227, 257]. The first fully combinatorial algorithm for the SFM has been described in [156], improved in [157], and recently improved again (without using the scaling method) in [160].

The time complexity of the fastest known general strongly polynomial algorithm for the SFM is $O(n^6 + n^5 L)$, where n is the number of variables and L is the time required to evaluate the function [227].

We now define submodularity in a more general setting. Recall that L is a *lattice* if L is a partially ordered set in which every pair of elements has a unique supremum (called *join*) and a unique infimum (called *meet*). For a finite lattice L and a pair of elements (a, b), we will denote the unique supremum of a and b by $\text{Max}(a, b)$ (or $a \vee b$), and the unique infimum of a and b by $\text{Min}(a, b)$ (or $a \wedge b$).

Definition 1.12 (Submodularity on lattices) Let D be a finite lattice-ordered set with the meet operation Min and the join operation Max. A function $\phi : D^n \to \overline{\mathbb{Q}}_{\geq 0}$

---

[8]Let $\mathcal{B}$ be an algorithm for the minimisation problem of f in the oracle value model. $\mathcal{B}$ is called *polynomial* if it runs in polynomial time. A polynomial algorithm $\mathcal{B}$ is called *strongly polynomial* if the running time does not depend on $M = \max f$. In other words, the number of elementary arithmetic operations and other operations is bounded by a polynomial in the size of the input. A polynomial algorithm $\mathcal{B}$ which does depend on M is called *weakly polynomial*. A polynomial algorithm $\mathcal{B}$ is called *combinatorial* if it does not employ the ellipsoid method. Finally, a combinatorial algorithm $\mathcal{B}$ is called *fully combinatorial* if it uses only oracle calls, additions, subtractions, and comparisons, but not multiplications and divisions, as fundamental operations.

is *submodular* if $\langle \text{Min}, \text{Max} \rangle \in \text{Mul}(\{\phi\})$, that is, for all n-tuples u, v,

$$\phi\big(\text{Min}(u, v)\big) + \phi\big(\text{Max}(u, v)\big) \leq \phi(u) + \phi(v). \tag{1.4}$$

Submodular functions on sets are equivalent to submodular functions on distributive lattices [258]. The minimisation problem of submodular functions on arbitrary lattices was first studied in [191].

An important and well-studied subproblem of SFM is the minimisation of submodular functions of bounded arity (SFM$_b$), also known as *locally-defined* submodular functions [73], or submodular functions with *succinct representation* [114]. In this scenario the submodular function to be minimised is defined by the sum of a collection of functions which each depend only on a bounded number of variables. Locally-defined optimisation problems of this kind, apart from being equivalent to VCSPs, occur in other contexts:

- In the context of *pseudo-Boolean optimisation*, such problems involve the minimisation of Boolean polynomials of bounded *degree* [34, 84].
- In the context of computer vision, such problems are often formulated as *Gibbs Energy Minimisation* problems [128] or *Markov Random Fields* (also known as *Conditional Random Fields*) [203, 267].

Definition 1.13 A language Γ defined over D is called *submodular* if $\Gamma \subseteq \text{Imp}(\langle \text{Min}, \text{Max} \rangle)$, where Min and Max are the meet and join operations with respect to some lattice order on D.

Cohen et al. have shown that VCSP instances with submodular constraints (with respect to some total order) over an arbitrary finite domain can be reduced to the SFM [67], and hence can be solved in polynomial time. This tractability result has since been generalised to a wider class[9] of valued constraints over arbitrary finite domains known as tournament pair valued constraints [66], and also to a class of valued constraints admitting complementary STP and MJN multimorphisms (more in Chap. 7) [188]. An alternative approach to solving VCSP instances with submodular constraints, based on linear programming, has been proposed in [73]. Recently, Thapper and Živný have shown that linear programming can be used to solve VCSPs with submodular constraints with respect to *any* lattice. This result will be discussed in Chap. 8.

The class of submodular languages with respect to some total order is the only nontrivial tractable class of optimisation problems in the dichotomy classification of Boolean VCSPs [67] (cf. Example 1.15 and Chap. 6), of finite-valued conservative VCSPs [188] (cf. Example 1.25 and Chap. 7), and of MAX-CSPs for both three-element domains [171] and arbitrary finite domains allowing constant constraints [100] (and hence also of conservative MAX-CSPs); cf. Example 1.16.

---

[9]The class of cost functions closed under a tournament pair multimorphism is more general than the class of submodular cost functions if the range of the cost functions includes infinite costs [66].

Part II
Expressive Power

Chapter 2
Expressibility of Valued Constraints

> *In mathematics you don't understand things. You just get used to them.*
> John von Neumann

2.1 Introduction

It has been known for some time that the expressive power of crisp constraints is determined by algebraic operations called polymorphisms. Moreover, there is a Galois connection between the set of crisp constraints and the set of operations. This connection has been successfully used in the last decade for complexity analysis of crisp constraint languages [42, 46].

This chapter presents a recently developed algebraic theory for the complexity of valued constraint languages [63, 68], which builds on [65, 274].

2.2 Results

In Sect. 2.3, we survey what is known about the expressive power of crisp constraints, describe the concept of the indicator problem and introduce the concept of a Galois connection. In Sect. 2.4, we show that the question of whether a given cost function is expressible over a finite language is decidable [65]. In Sect. 2.5, we present dual results for fractional operations [274]. Putting results from Sects. 2.4 and 2.5 together, Sect. 2.6 presents a Galois connection between sets of valued constraints and sets of algebraic closure operations [68].

2.3 Indicator Problem

In this section, we discuss the well-known result that the expressive power of crisp constraints is characterised by certain algebraic operations called polymorphisms. We present the construction of an indicator problem, a universal construction for determining whether a given relation is expressible over a crisp constraint language,

S. Živný, *The Complexity of Valued Constraint Satisfaction Problems*,
Cognitive Technologies, DOI 10.1007/978-3-642-33974-5_2,
© Springer-Verlag Berlin Heidelberg 2012

and also for determining all polymorphisms of a crisp constraint language. Finally, we show that there is a Galois connection between the set of relations and the set of operations.

Recall Theorem 1.2, which states that the expressive power of a crisp constraint language is fully characterised by its polymorphisms [27, 127, 163]. In other words, for a relation R and a crisp constraint language Γ, the following holds:

$$R \in \langle \Gamma \rangle \quad \Leftrightarrow \quad \mathsf{Pol}(\Gamma) \subseteq \mathsf{Pol}(\{R\}).$$

Remark 2.1 The "$\Rightarrow$" implication follows easily from the fact that expressibility preserves polymorphisms.

This result was obtained by showing that, for any crisp language (that is, set of relations), there is a *universal construction* that can be used to determine whether a relation is expressible in that language, as we now demonstrate.

Definition 2.1 (Indicator Problem)   Let Γ be a crisp constraint language over D. For any natural number n, we define the *indicator problem* for Γ of order n as the CSP instance $\mathcal{IP}(\Gamma, n)$ with the set of variables D^n, each with domain D, and constraints $\{C_i\}_{1 \le i \le q}$, where $q = \sum_{R \in \Gamma} |R|^n$. For each $R \in \Gamma$, and for each sequence $t_1, t_2, \ldots, t_n$ of tuples from R, there is a constraint $C_i = \langle s_i, R \rangle$ with $s_i = \langle v_1, v_2, \ldots, v_m \rangle$, where m is the arity of R, and $v_j = \langle t_1[j], t_2[j], \ldots, t_n[j] \rangle$, $1 \le j \le m$.

Note that, for any crisp constraint language Γ over D, $\mathcal{IP}(\Gamma, n)$ has $|D|^n$ variables, and each corresponds to an n-tuple over D. A concrete example of an indicator problem is given below, and more examples can be found in [164, 166].

Observation 2.1   It is not hard to see from Definitions 2.1 and 1.6 that the solutions to $\mathcal{IP}(\Gamma, n)$ are the polymorphisms of Γ of arity n [165].

Combining Observation 2.1 with Theorem 1.2 gives:

Corollary 2.1   *Let Γ be a crisp constraint language over D. Furthermore, let $R = \{t_1, t_2, \ldots, t_n\}$ be a relation over D of arity m. Then R is expressible over Γ, $R \in \langle \Gamma \rangle$, if, and only if, R is equal to the solutions to $\mathcal{IP}(\Gamma, n)$ restricted to the variables $v_1, v_2, \ldots, v_m$, where $v_j = \langle t_1[j], t_2[j], \ldots, t_n[j] \rangle$, $1 \le j \le m$.*

Note that the choice of variables $v_1, v_2, \ldots, v_m$ might not be unique as different orderings of the tuples of R can result in different lists. We sketch the proof of Corollary 2.1 since it contains the idea behind the proof of Theorem 1.2.

Proof (Sketch)   From Definition 2.1, if R is equal to the solutions to $\mathcal{IP}(\Gamma, n)$ restricted to some subset of variables, then R is expressible over Γ. On the other hand, assume that $R \in \langle \Gamma \rangle$, and denote by $\bar{R}$ the set of solutions to $\mathcal{IP}(\Gamma, n)$ restricted to

Fig. 2.1 $\mathcal{IP}(\Gamma, 3)$
(Example 2.1)

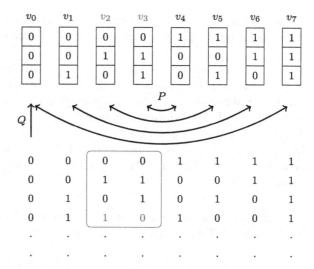

the variables $v_1, v_2, \ldots, v_m$. It is enough to show that $R = \bar{R}$. By Observation 1.1, all projections are polymorphisms of all relations. Hence, by Observation 2.1, all projections of arity n are solutions of $\mathcal{IP}(\Gamma, n)$. Therefore, $R \subseteq \bar{R}$ from the choice of variables $v_1, v_2, \ldots, v_m$. If $R \neq \bar{R}$, then there must be a solution s to $\mathcal{IP}(\Gamma, n)$ whose restriction to $v_1, v_2, \ldots, v_m$ is not contained in R. By Observation 2.1, all solutions to $\mathcal{IP}(\Gamma, n)$ are polymorphisms of Γ, and so is s. But by Remark 2.1, the polymorphism s should be a polymorphism of R, which is a contradiction if $R \neq \bar{R}$. $\square$

Remark 2.2 Richard Gault implemented a solver called POLYANNA[1] for the indicator problem [126]. An interesting research problem is to investigate various symmetries in the indicator problem and try to make use of them in order to solve instances of the indicator problem more efficiently. However, there is a known worst-case lower bound on the size of any expressibility gadget [270].

Example 2.1 Let $\Gamma = \{P, Q\}$ be a constraint language over $D = \{0, 1\}$, where $P = \{\langle 0, 1 \rangle, \langle 1, 0 \rangle\}$ and $Q = \{\langle 0 \rangle\}$. Given relation $R = \{\langle 0, 0 \rangle, \langle 0, 1 \rangle, \langle 1, 1 \rangle\}$, the task is to determine whether R is expressible over Γ.

Since R consists of three tuples, we construct the indicator problem $\mathcal{IP}(\Gamma, 3)$ of order 3. The variables of $\mathcal{IP}(\Gamma, 3)$ are all 3-tuples over D. There are eight 3-tuples over D, and we denote them by v_0 to v_7 (cf. Fig. 2.1). Since the relation P is binary, any two variables v_i and v_j, which represent the tuples $\langle v_{i1}, v_{i2}, v_{i3} \rangle$ and $\langle v_{j1}, v_{j2}, v_{j3} \rangle$, respectively, are constrained by P if, and only if, all three tuples $\langle v_{i1}, v_{j1} \rangle$, $\langle v_{i2}, v_{j2} \rangle$, and $\langle v_{i3}, v_{j3} \rangle$ belong to P. In our case the following pairs of variables are constrained by P: $\langle v_0, v_7 \rangle$, $\langle v_1, v_6 \rangle$, $\langle v_2, v_5 \rangle$, and $\langle v_3, v_4 \rangle$. The relation

---

[1]http://www.cs.ox.ac.uk/activities/constraints/software/index.html.

Q is unary and consists of just one tuple $\langle 0 \rangle$. Therefore, only the variable v_0, which represents the tuple $\langle 0, 0, 0 \rangle$, is constrained by Q. The construction is illustrated in Fig. 2.1.

Now consider the tuples represented by the variables v_2 and v_3. These are $\langle 0, 1, 0 \rangle$ and $\langle 0, 1, 1 \rangle$, respectively. If you take these two tuples as columns of a matrix, then the rows of this matrix contain precisely the tuples from R, that is, $\langle 0, 0 \rangle$, $\langle 0, 1 \rangle$, and $\langle 1, 1 \rangle$. However, projecting the solutions to $\mathcal{IP}(\Gamma, 3)$ onto variables v_2 and v_3 does not give R, as the tuple $\langle 1, 0 \rangle$ does not belong to R. (Some of the solutions to $\mathcal{IP}(\Gamma, 3)$ are shown in Fig. 2.1.) Hence R is not expressible over Γ. (Note that we could also obtain the same result by choosing, for instance, variables v_4 and v_6.)

Recall from Notation 1.8 that, for a crisp constraint language Γ, we denote by $\mathsf{Pol}(\Gamma)$ the set of all polymorphisms of Γ, that is,

$$\mathsf{Pol}(\Gamma) = \{ f \mid \forall R \in \Gamma, f \text{ is a polymorphism of } R \}.$$

Notation 2.1 For a set of operations O, we use $\mathsf{Inv}(O)$ to denote the set of relations having all operations in O as polymorphisms, that is,

$$\mathsf{Inv}(\Gamma) = \{ R \mid \forall f \in O, f \text{ is a polymorphism of } R \}.$$

Observation 2.2 The result of Theorem 1.2 can be equivalently stated as follows: for any crisp constraint language Γ, it holds that $\langle \Gamma \rangle = \mathsf{Inv}(\mathsf{Pol}(\Gamma))$.

Definition 2.2 (Galois connection [101]) A *Galois connection* between two sets A and B is a pair $\langle F, G \rangle$ of mappings between the power sets $\mathcal{P}(A)$ and $\mathcal{P}(B)$, $F : \mathcal{P}(A) \rightarrow \mathcal{P}(B)$ and $G : \mathcal{P}(B) \rightarrow \mathcal{P}(B)$, such that for all $X, X' \subseteq A$ and all $Y, Y' \subseteq B$ the following conditions are satisfied: $X \subseteq X' \Rightarrow F(X) \supseteq F(X')$ and $Y \subseteq Y' \Rightarrow G(Y) \supseteq G(Y')$; $X \subseteq G(F(X))$ and $Y \subseteq F(G(Y))$.

Notation 2.2 For any finite domain D, we denote by $\mathbf{R}_D$ the set of all relations over D, and we denote by $\mathbf{O}_D$ the set of all operations over D.

The following easy result shows that the $\mathsf{Pol}(\cdot)$ and $\mathsf{Inv}(\cdot)$ operators give rise to an instance of a *Galois connection* between $\mathbf{R}_D$ and $\mathbf{O}_D$ for any finite domain D.

Proposition 2.1 ([234]) *If Γ is a set of relations over D and O is a set of operations over D, then*

1. $O_1 \subseteq O_2 \subseteq \mathbf{O}_D \Rightarrow \mathsf{Inv}(O_1) \supseteq \mathsf{Inv}(O_2)$.
2. $\Gamma_1 \subseteq \Gamma_2 \subseteq \mathbf{R}_D \Rightarrow \mathsf{Pol}(\Gamma_1) \supseteq \mathsf{Pol}(\Gamma_2)$.
3. $\Gamma \subseteq \mathsf{Inv}(\mathsf{Pol}(\Gamma))$.
4. $O \subseteq \mathsf{Pol}(\mathsf{Inv}(O))$.

In order for a Galois connection to be interesting, one should be able to characterise the closures under the two operators. In particular, we are interested in the closure of a set of relations and in the closure of a set of operations under the $\mathsf{Pol}(\cdot)$ and $\mathsf{Inv}(\cdot)$ operators.

Recall that for a crisp constraint language $\Gamma \subseteq \mathbf{R}_D$, we denote by $\langle \Gamma \rangle$ the set of relations that are expressible over Γ. The set $\langle \Gamma \rangle$ is also known as a *relational clone* [234]. For a set of operations $F \subseteq \mathbf{O}_D$, we denote by $\langle F \rangle$ the set of operations from F closed under superposition (also known as composition)[2] and containing all projections (cf. Observation 1.1). The set $\langle F \rangle$ is known as a *clone of operations*, or just a *clone* [234].

A characterisation of this Galois connection for finite sets D is given by the following two theorems, originally obtained for sets of relations [27, 127].

Theorem 2.1 *For any finite set D, and any $\Gamma \subseteq \mathbf{R}_D$, $\mathsf{Inv}(\mathsf{Pol}(\Gamma)) = \langle \Gamma \rangle$.*

Theorem 2.2 *For any finite set D, and any $F \subseteq \mathbf{O}_D$, $\mathsf{Pol}(\mathsf{Inv}(F)) = \langle F \rangle$.*

The situation is summarised in Fig. 2.2. As with any Galois connection [31], this means that there is a one-to-one correspondence between clones and relational clones. This result shows that the expressive power of any crisp constraint language Γ on a finite set D corresponds to a particular clone of operations on D. Hence the search for tractable crisp constraint languages corresponds to a search for suitable clones of operations [42, 163]. This key observation paved the way for applying deep results from universal algebra in the search for tractable constraint languages [11, 13, 14, 16, 38, 39, 41, 45].

Post completely described the lattice of relational clones and clones over a two-element domain [235]. This description has been heavily used recently to obtain dichotomy complexity classifications of various classes of problems in computer science and artificial intelligence that can be modelled over Boolean domains. For more details, see [28, 29]. More on clone theory can be found in [101, 234].

2.4 Weighted Indicator Problem

In this section, we show that there is also a universal construction to determine whether a given cost function is expressible over a valued constraint language. We briefly describe the result that the expressive power of valued constraints is determined by certain algebraic operations called fractional polymorphisms.

Consider the following problem: given a cost function ϕ of arity m over a domain D, is ϕ expressible over a valued constraint language Γ? We show that this problem is decidable for every finite Γ. First we prove an upper bound on the number of extra variables needed to express ϕ over Γ.

---

[2]Let $f : D^k \to D$ and $g_1, \ldots, g_k : D^\ell \to D$. The superposition of f and $g_1, \ldots, g_k$ is the ℓ-ary operation $f[g_1, \ldots, g_k](x_1, \ldots, x_\ell) = f(g_1(x_1, \ldots, x_\ell), \ldots, g_k(x_1, \ldots, x_l))$.

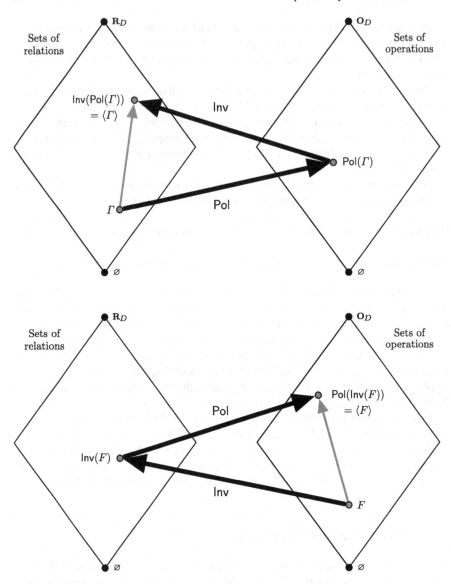

Fig. 2.2 Galois connection between $\mathbf{R}_D$ and $\mathbf{O}_D$

Proposition 2.2 ([65]) *If a cost function $\phi : D^m \to \overline{\mathbb{Q}}_{\geq 0}$ is expressible over Γ, then ϕ is expressible over Γ using at most $|D|^{|D|^m}$ hidden variables.*

Proof If $\phi \in \langle \Gamma \rangle$, then by Definition 1.4, there is a gadget $\langle \mathcal{P}, \mathbf{v} \rangle$, where $\mathbf{v} = \langle v_1, \ldots, v_m \rangle$, for expressing ϕ over Γ. For the gadget $\langle \mathcal{P}, \mathbf{v} \rangle$ to express ϕ, it has to define ϕ on each of the $|D|^m$ different assignments to $\mathbf{v}$. Let each of these $|D|^m$ assignments be extended to a complete assignment to all variables of $\mathcal{P}$ (including hidden variables) in a way that minimises the total cost. For each hidden variable v of $\langle \mathcal{P}, \mathbf{v} \rangle$, we can use the list of $|D|^m$ values assigned to v by these complete assignments to label the variable v. If there are more than $|D|^{|D|^m}$ hidden variables, then two of them will receive the same label. However, this implies that one of the two is redundant, as all constraints involving that variable can replace it with the other variable without changing the overall cost. Hence we require at most $|D|^{|D|^m}$ distinct hidden variables to express ϕ.   □

From this bound on the number of extra variables in a gadget for ϕ over Γ we obtain a decidability result. The idea is to try all possible constraints on all possible subsets of variables, and use linear programming to determine whether there is a combination of these constraints which works.

Theorem 2.3 ([65]) *For a given finite valued constraint language Γ, and a cost function ϕ defined over D, the question of whether ϕ is expressible over Γ is decidable.*

Proof (Sketch) In order to simplify the presentation, we assume that ϕ is a finite-valued cost function. We show how to determine whether there is a gadget for ϕ over Γ, that is, whether there is a VCSP(Γ) instance $\mathcal{P} = \langle V, D, \mathcal{C} \rangle$ and a tuple of variables $\mathbf{v}$ such that $\phi = \pi_{\mathbf{v}}(\mathcal{P})$. By Proposition 2.2, $\mathcal{P}$ has at most $K = |D|^{|D|^m}$ extra variables, where m is the arity of ϕ. Let V be the set of K variables, each associated with a different $|D|^m$-tuple of values from D. Let E be the $|D|^m \times K$ matrix whose columns are all possible $|D|^m$-tuples of values from D (or equivalently, variables from V). Observe that there is a $|D|^m \times m$ submatrix E' of E consisting of m columns of E such that the rows of E' correspond to all possible m-tuples of values from D. We let $\mathbf{v}$ be the list of variables corresponding to the columns of E'.

Let $\mathcal{A}$ be the set of all assignments of values from D to the variables from V. Clearly, $|\mathcal{A}| = |D|^K$. We choose $|D|^m$ assignments from $\mathcal{A}$ that correspond to the rows of the matrix E and denote them $\mathcal{A}'$.

Let $\rho \in \Gamma$ be a cost function of arity k. For an assignment $s \in \mathcal{A}$ and a list of variables $\mathbf{u}$ of length k, recall from Definition 1.1 that we denote by $\rho(s(\mathbf{u}))$ the value of ρ on the list of variables $\mathbf{u}$ assigned by s.

The idea is that if ϕ is expressible over Γ, then there is a set of constraints $\mathcal{C}$ such that $\phi = \pi_{\mathbf{v}}(\mathcal{P})$, where $\mathcal{P} = \langle V, D, \mathcal{C} \rangle$. It remains to show what the set of constraints $\mathcal{C}$ is. And this is where linear programming plays its crucial role.

Let $\mathcal{C}$ be the list of all possible constraints from Γ applied to variables from V. In other words, $\mathcal{C} = \langle C_1, \ldots, C_q \rangle$ is an arbitrary but fixed order of the following finite set:

$$\{ \langle \mathbf{u}, \rho \rangle \mid \rho \in \Gamma \text{ of arity } k, \text{ and } \mathbf{u} \text{ is a list of } k \text{ variables from } V \}.$$

We write $C_i = \langle \mathbf{u}_i, \rho_i \rangle$. Clearly,

$$q = \sum_{\rho \in \Gamma \text{ of arity } k} K^k.$$

We define a system of linear equations and inequalities as follows.

For each $s \in \mathcal{A} \setminus \mathcal{A}'$,

$$\sum_{i=1}^{q} x_i \rho_i \big(s(\mathbf{u}_i)\big) \geq \phi\big(s(\mathbf{v})\big) + x_0.$$

For each $s \in \mathcal{A}'$,

$$\sum_{i=1}^{q} x_i \rho_i \big(s(\mathbf{u}_i)\big) = \phi\big(s(\mathbf{v})\big) + x_0.$$

Note that the variable x_i represents whether the constraint C_i is used in the gadget $\mathcal{P}$: if $x_i = 0$, then the constraint C_i is not used; if $x_i > 0$, then x_i gives the multiplicity of the constraint C_i. The variable x_0 represents an additive constant up to which the gadget expresses ϕ.

From the construction of the system, ϕ is expressible over Γ if, and only if, there is a nonnegative solution to this system, which is decidable [256]; see also [7]. □

Remark 2.3 Theorem 2.3 can be extended from finite-valued cost functions to general-valued cost functions [65]. The construction sketched above is known as the *weighted indicator problem*.

Example 2.2 Let $\Gamma = \{\mu, \psi\}$ be the valued constraint language consisting of two cost functions defined over the Boolean domain $D = \{0, 1\}$ as follows:

$$\mu(x) \stackrel{\text{def}}{=} \begin{cases} 0 & \text{if } x = 0, \\ 1 & \text{if } x = 1, \end{cases}$$

and

$$\psi(x, y) \stackrel{\text{def}}{=} \begin{cases} -1 & \text{if } x = y = 1, \\ 0 & \text{otherwise.} \end{cases}$$

Let ϕ be the ternary cost function defined as follows:

$$\phi(x, y, z) \stackrel{\text{def}}{=} \begin{cases} -1 & \text{if } x = y = z = 1, \\ 0 & \text{otherwise.} \end{cases}$$

The question is whether ϕ is expressible over Γ, that is, whether $\phi \in \langle \Gamma \rangle$. In order to answer this question, we build an instance of the weighted indicator problem as described in the sketched proof of Theorem 2.3. The arity m of ϕ is 3, and hence if ϕ is expressible over Γ, then ϕ is expressible with at most $K = |D|^{|D|^m} = 2^{2^3} = 2^8 = 256$ variables, by Proposition 2.2. Each variable is uniquely identified by an 8-tuple of values from $\{0, 1\}$. We denote by V the set of all such variables with the domain $\{0, 1\}$.

We denote by $\mathbf{v} = \langle v_1, v_2, v_3 \rangle$ the list of three variables, whose corresponding 8-tuples are $t_1 = \langle 0, 0, 0, 0, 1, 1, 1, 1 \rangle$, $t_2 = \langle 0, 0, 1, 1, 0, 0, 1, 1 \rangle$, and $t_3 = \langle 0, 1, 0, 1, 0, 1, 0, 1 \rangle$ respectively. Consider the matrix whose columns are tuples t_1,

t_2, and t_3. The rows of this matrix are all possible 3-tuples over $\{0, 1\}$. The intuition is that we try to find a gadget $\mathcal{P}$ for ϕ over Γ that expresses ϕ on the variables v_1, v_2, and v_3, that is, $\phi = \pi_{\langle v_1, v_2, v_3 \rangle}(\mathcal{P})$.

Let E be an 8×256 matrix whose columns are the tuples corresponding to variables from V in some fixed order.

We denote by $\mathcal{A}'$ the set of eight assignments of variables from V that are defined by the rows of the matrix E. The intuition is that for every possible assignment s of the variables v_1, v_2, and v_3, we are looking for an assignment s' in $\mathcal{A}'$ which agrees with s (on v_1, v_2, and v_3) and the cost of s' is equal to $\phi(v_1, v_2, v_3)$ (up to an additive constant). We denote by $\mathcal{A}$ all assignments of variables from V. Clearly, $|\mathcal{A}| = 2^{256}$.

Now we want to add all possible constraints involving cost functions from Γ. The unary cost function μ can be applied to any of the 256 variables. The binary cost function ψ can be applied to any pair of (not necessarily distinct) variables. Since ψ is symmetric, this gives $\binom{256}{2} + 256$ constraints. In total, we get $2 * 256 + \binom{256}{2} = 33,152$ constraints. Hence we have 33,152 variables x_i that represent whether the ith constraint is used ($x_i > 0$) or not ($x_i = 0$); in the former case, the value of x_i represents the multiplicity of the ith constraint. We then can build a system of linear equations and inequalities as described in the sketch of the proof of Theorem 2.3.

In this particular case, it is not difficult to find a solution to the system of linear equations and inequalities described above. Let y be the variable corresponding to the 8-tuple $\langle 0, 0, 0, 0, 0, 0, 0, 1 \rangle$. We claim that assigning the value 2 to the constraint $\langle \langle y \rangle, \mu \rangle$ (represented by a variable in our system), assigning the value 1 to the constraints $\langle \langle y, x_1 \rangle, \psi \rangle$, $\langle \langle y, x_2 \rangle, \psi \rangle$, and $\langle \langle y, x_3 \rangle, \psi \rangle$, and finally assigning the value 0 to the additive constant $x_0 = 0$ and all other variables is a solution to our system. For any assignment of the variables v_1, v_2, and v_3, setting y to 0 results in total cost 0. If all v_1, v_2, and v_3 are assigned 1, setting y to 1 results in total cost -1. For any other assignment of v_1, v_2, and v_3, setting y to 1 results in total cost ≥ 0. This corresponds exactly to the definition of the cost function ϕ. This solution gives a gadget for expressing ϕ over Γ using only one extra variable.

Recall Theorem 1.3, which states that the expressive power of a valued constraint language satisfying certain conditions is fully characterised by its polymorphisms and fractional polymorphisms [65]. In other words, for a cost function ϕ and a valued constraint language Γ, such that Γ contains a constant function and is closed under scaling and the feasibility operator, the following holds:

$$\phi \in \langle \Gamma \rangle \quad \Leftrightarrow \quad \mathsf{Pol}(\Gamma) \subseteq \mathsf{Pol}(\{\phi\}) \wedge \mathsf{fPol}(\Gamma) \subseteq \mathsf{fPol}(\{\phi\}).$$

Remark 2.4 The "$\Rightarrow$" implication follows easily from the fact that expressibility preserves polymorphisms and fractional polymorphisms [65]; see also [67].

Remark 2.5 We remark on the assumptions of Theorem 1.3. Notice that it is not a restrictive assumption that every valued constraint language Γ contains a constant function and is closed under scaling. In practice, this corresponds to adding a finite constant that does not alter the relative costs, and to taking more copies of the same constraint. Therefore, this does not change the complexity of solving VCSP instances over Γ.

We now discuss why it is necessary to assume that Γ is closed under the feasibility operator (or, equivalently, closed under scaling by 0; cf. Remark 1.10) in order to prove equivalence in Theorem 1.3. If $\mathcal{F}$ is a fractional polymorphism of Γ, then $\mathcal{F}$ is also a fractional polymorphism of $\text{Feas}(\Gamma)$. And clearly, any polymorphism of $\text{Feas}(\Gamma)$ is a polymorphism of Γ. Hence for any valued constraint language Γ, $\text{Pol}(\Gamma) \subseteq \text{Pol}(\text{Feas}(\Gamma))$ and $\text{fPol}(\Gamma) \subseteq \text{fPol}(\text{Feas}(\Gamma))$. But this is true for any Γ independently of whether or not $\text{Feas}(\Gamma) \in \langle \Gamma \rangle$; so every valued constraint language Γ satisfies the right-hand side of the equivalence in Theorem 1.3 for $\text{Feas}(\Gamma)$ (that is, $\text{Pol}(\Gamma) \subseteq \text{Pol}(\text{Feas}(\Gamma))$ and $\text{fPol}(\Gamma) \subseteq \text{fPol}(\text{Feas}(\Gamma)))$, but not every valued constraint language Γ satisfies $\text{Feas}(\Gamma) \in \langle \Gamma \rangle$.

Fractional polymorphisms on their own characterise the expressive power of finite-valued cost functions and, as shown in Theorem 1.2, polymorphisms on their own characterise the expressive power of crisp cost functions that take only zero and infinite costs.

The proof of Theorem 1.3 given in [65] is based on an application of Farkas' Lemma [256] and uses the concept of the weighted indicator problem. For a given ϕ and Γ, as sketched above, a system of linear equations and inequalities is built such that it has a solution if, and only if, ϕ is expressible over Γ. If this system does not have a solution, then Farkas' Lemma guarantees a certificate of unsolvability. Cohen et al. have shown that the certificate of unsolvability can be turned into a fractional polymorphism $\mathcal{F}$ such that $\mathcal{F} \in \text{fPol}(\Gamma)$, but $\mathcal{F} \notin \text{fPol}(\{\phi\})$ [65].

2.5 Fractional Clones

In this section, we consider the dual question to the one considered in Sect. 2.4: given a finite set Ω of fractional operations over a fixed domain D and a fractional operation $\mathcal{F}$ defined over D, determine whether or not $\mathcal{F}$ belongs to $\text{fPol}(\text{Imp}(\Omega))$. We call this problem the *fractional clone membership* problem.

Using linear programming, we show that this problem is decidable.

Theorem 2.4 *The fractional clone membership problem is decidable.*

Proof Let $\mathcal{F}$ be a k-ary fractional operation $\{\langle r_1, f_1 \rangle, \ldots, \langle r_n, f_n \rangle\}$ such that each f_i is a distinct function from D^k to D, each r_i is a positive rational number, and $\sum_{i=1}^{n} r_i = k$. Let $\Omega = \{\mathcal{F}_1, \ldots, \mathcal{F}_q\}$.

Now $\mathcal{F} \notin \text{fPol}(\text{Imp}(\Omega))$ if, and only if, there is a finite-valued cost function ϕ such that $\mathcal{F}_i \in \text{fPol}(\{\phi\})$ for every $1 \leq i \leq q$, but $\mathcal{F} \notin \text{fPol}(\{\phi\})$.

First we show that if there is such a ϕ (we call it a *separating cost function*), then there is a ϕ of arity at most $m = |D|^k$. Assume that ϕ is of arity strictly larger than m. As there are exactly m different k-tuples over D, any tableau (recall Fig. 1.5) showing that $\mathcal{F}_i \in \mathsf{fPol}(\{\phi\})$, $1 \leq i \leq q$, and $\mathcal{F} \notin \mathsf{fPol}(\{\phi\})$ has at least one column that occurs twice. However, this column can be removed and the arity of ϕ decreased by 1 by identifying the two arguments corresponding to the two columns. Clearly, if there is a separating cost function ϕ of arity strictly smaller than m, then there is a separating cost function of arity exactly m: we just add dummy variables. Hence we can assume that ϕ is of arity exactly m.

Now we can turn the question of the existence of a separating cost function into a system of linear inequalities. We are looking for $|D|^m$ values (costs of ϕ on all possible assignments) such that for all $1 \leq i \leq q$, $\mathcal{F}_i \in \mathsf{fPol}(\{\phi\})$, and $\mathcal{F} \notin \mathsf{fPol}(\{\phi\})\}$. But this is easy as showing that $\mathcal{F}_i \in \mathsf{fPol}(\{\phi\})$ is just a question of satisfying a system of linear inequalities for all possible tableaux, by Definition 1.10. Similarly, $\mathcal{F} \notin \mathsf{fPol}(\{\phi\})$ can be expressed as one linear inequality corresponding to the tableau with m different k-tuples over D and the inequality sign the opposite of that in Definition 1.10. This finishes the proof, as the question of whether there is a solution to a system of linear inequalities is decidable [256]. □

The following example illustrates the technique described in the proof of Theorem 2.4.

Example 2.3 Let $\mathcal{F}_1 = \{\langle 1, \mathsf{Min} \rangle, \langle 1, \mathsf{Max} \rangle\}$, $\mathcal{F} = \{\langle 2, \mathsf{Max} \rangle\}$, and $D = \{0, 1\}$. In order to determine whether $\mathcal{F} \in \mathsf{fPol}(\mathsf{Imp}(\mathcal{F}_1))$, we build a system of linear inequalities as in the proof of Theorem 2.4. We look for a separating cost function ϕ of arity $m = |D|^2 = 4$. Hence we have the $|D|^4 = 16$ variables $x_{0000}, x_{0001}, \ldots, x_{1111}$ corresponding to the values of ϕ on 16 different assignments. There are $16 * 16 = 256$ inequalities that make sure that $\{\langle 1, \mathsf{Min} \rangle, \langle 1, \mathsf{Max} \rangle\} \in \mathsf{fPol}(\{\phi\})$:

$$x_{ijkl} + x_{mnop} \geq x_{abcd} + x_{uvyz},$$

where $a = \min(i, m)$, $b = \min(j, n)$, $c = \min(k, o)$, $d = \min(l, p)$, and $u = \max(i, m)$, $v = \max(j, n)$, $y = \max(k, o)$, $z = \max(l, p)$.

Another inequality makes sure that $\{\langle 2, \mathsf{Max} \rangle\} \notin \mathsf{fPol}(\{\phi\})$. According to the proof of Theorem 2.4, the tableau consists of four 2-tuples over $\{0, 1\}$. Hence, the required inequality is

$$x_{0011} + x_{0101} < x_{0111} + x_{0111},$$

where $T = \binom{0011}{0101}$ on the left-hand side corresponds to four different 2-tuples (columnwise), and $\binom{0111}{0111}$ on the right-hand side is the application of O to T.

One solution to this system is $x_{00..} = 0$ and $x_{01..} = x_{10..} = x_{11..} = 1$. Notice that ϕ is effectively binary as it only depends on its first two arguments: $\phi(x, y, \cdot, \cdot) = 0$ if $x = y = 0$, and $\phi(x, y, \cdot, \cdot) = 1$ otherwise. It is straightforward to check that this is indeed a solution to the system; that is, $\{\langle 1, \mathsf{Min} \rangle, \langle 1, \mathsf{Max} \rangle\} \in \mathsf{fPol}(\{\phi\})$, but $\{\langle 2, \mathsf{Max} \rangle\} \notin \mathsf{fPol}(\{\phi\})$.

2.6 Galois Theory

We have seen in Theorem 1.3 that the expressive power of languages is determined by the polymorphisms and fractional polymorphisms of the language. However, in order to obtain a Galois connection similar to the one presented for crisp languages in Sect. 2.3, we will have to generalise slightly fractional polymorphisms. The main idea is the following: in the tableau in Fig. 1.5, the upper part corresponds to projections and the bottom part to operations. We relax the definition so that (in some cases) both parts can be arbitrary operations.

Recall that a clone of operations, C, is a set of operations on some fixed set D that contains all projections and is closed under superposition. The k-ary operations in a clone C will be denoted by $C^{(k)}$.

Definition 2.3 (Weighting) We define a k-ary *weighting* of a clone C to be a function $\omega : C^{(k)} \to \mathbb{Q}$ such that $\omega(f) < 0$ only if f is a projection and

$$\sum_{f \in C^{(k)}} \omega(f) = 0.$$

We denote by $\mathbf{W}_C$ the set of all possible weightings of C and by $\mathbf{W}_C^{(k)}$ the set of k-ary weightings of C.

For any weighting ω, we denote by $\mathbf{dom}(\omega)$ the set of operations on which ω is defined.

Since a weighting is simply a rational-valued function satisfying certain inequalities it can be scaled by any nonnegative rational to obtain a new weighting. Similarly, any two weightings of the same clone of the same arity can be added to obtain a new weighting of that clone.

The notion of superposition for operations can also be extended to weightings in a natural way, as follows.

Definition 2.4 For any clone C, any $\omega \in \mathbf{W}_C^{(k)}$ and any $g_1, g_2, \ldots, g_k \in C^{(l)}$, we define the *superposition* of ω and $g_1, \ldots, g_k$, to be the weighting $\omega[g_1, \ldots, g_k] \in \mathbf{W}_C^{(l)}$ defined by

$$\omega[g_1, \ldots, g_k](f') \stackrel{\text{def}}{=} \sum_{\substack{f \in C^{(k)} \\ f' = f[g_1, \ldots, g_k]}} \omega(f). \tag{2.1}$$

Example 2.4 Let C be a clone on some ordered set D and let Max and Min be binary maximum and minimum operations contained in C. Note that $C^{(4)}$ contains operations such as $\text{Max}[e_i^{(4)}, e_j^{(4)}]$, which returns the maximum of the ith and jth argument values. Operations of this form will be denoted by $\text{Max}(x_i, x_j)$.

Let ω be the 4-ary weighting of C given by

$$\omega(f) \overset{\text{def}}{=} \begin{cases} -1 & \text{if } f \text{ is a projection, i.e., } f \in \{e_1^{(4)}, e_2^{(4)}, e_3^{(4)}, e_4^{(4)}\}, \\ +1 & \text{if } f \in \{\text{Max}(x_1, x_2), \text{Min}(x_1, x_2), \text{Max}(x_3, x_4), \text{Min}(x_3, x_4)\}, \\ 0 & \text{otherwise}, \end{cases}$$

and let

$$\langle g_1, g_2, g_3, g_4 \rangle = \langle e_1^{(3)}, e_2^{(3)}, e_3^{(3)}, \text{Max}(x_1, x_2) \rangle.$$

Then, by Definition 2.4 we have

$$\omega[g_1, g_2, g_3, g_4](f) = \begin{cases} -1 & \text{if } f \text{ is a projection, i.e., } f \in \{e_1^{(3)}, e_2^{(3)}, e_3^{(3)}\}, \\ +1 & \text{if } f \in \left\{ \begin{array}{l} \text{Max}(x_1, x_2, x_3), \text{Min}(x_1, x_2), \\ \text{Min}(x_3, \text{Max}(x_1, x_2)) \end{array} \right\}, \\ 0 & \text{otherwise}. \end{cases}$$

Note that $\omega[g_1, g_2, g_3, g_4]$ satisfies the conditions of Definition 2.3 and hence is a weighting of C.

Example 2.5 Let C and ω be the same as in Example 2.4, but now consider

$$\langle g_1', g_2', g_3', g_4' \rangle = \langle e_1^{(4)}, \text{Max}(x_2, x_3), \text{Min}(x_2, x_3), e_4^{(4)} \rangle.$$

By Definition 2.4 we have

$$\omega[g_1', g_2', g_3', g_4'](f) = \begin{cases} -1 & \text{if } f \in \{e_1^{(4)}, \text{Max}(x_2, x_3), \text{Min}(x_2, x_3), e_4^{(4)}\}, \\ +1 & \text{if } f \in \left\{ \begin{array}{l} \text{Max}(x_1, x_2, x_3), \text{Min}(x_2, x_3, x_4), \\ \text{Min}(x_1, \text{Max}(x_2, x_3)), \\ \text{Max}(\text{Min}(x_2, x_3), x_4) \end{array} \right\}, \\ 0 & \text{otherwise}. \end{cases}$$

Note that $\omega[g_1', g_2', g_3', g_4']$ does not satisfy the conditions of Definition 2.3 because, for example, we have $\omega[g_1', g_2', g_3', g_4'](f) < 0$ when $f = \text{Max}(x_2, x_3)$, which is not a projection. Hence $\omega[g_1', g_2', g_3', g_4']$ is not a valid weighting of C.

Definition 2.5 If the result of a superposition is a valid weighting, then that superposition will be called a *proper* superposition.

Remark 2.6 The superposition of a projection operation and other projection operations is always a projection operation. So, by Definition 2.4, for any clone C and any $\omega \in \mathbf{W}_C^{(k)}$, if $g_1, \ldots, g_k \in C^{(l)}$ are projections, then the function $\omega[g_1, \ldots, g_k]$ can take negative values only on projections, and hence is a valid weighting. This means that a superposition with any list of projections is always a proper superposition.

We are now ready to define *weighted clones*.

Definition 2.6 A *weighted clone*, W, is a nonempty set of weightings of some fixed clone C that is closed under nonnegative scaling, addition of weightings of equal arity, and proper superposition with operations from C. The clone C is called the *support* of W.

Example 2.6 For any clone, C, the set $\mathbf{W}_C$ containing all possible weightings of C is a weighted clone with support C.

Example 2.7 For any clone, C, the set $\mathbf{W}_C^0$ containing all *zero-valued* weightings of C is a weighted clone with support C.

We now establish a link between weightings and cost functions, which will allow us to link weighted clones and languages closed under expressibility.

Definition 2.7 (Weighted Polymorphism) Let $\phi : D^m \to \overline{\mathbb{Q}}_{\geq 0}$ be an m-ary cost function on some set D and let ω be a k-ary weighting of some clone of operations C on the set D. We say that ω is a *weighted polymorphism* of ϕ if, for any $\mathbf{x}_1, \ldots, \mathbf{x}_k \in D^r$ such that $\phi(\mathbf{x}_i) < \infty$ for $i = 1, \ldots, k$, we have

$$\sum_{f \in C^{(k)}} \omega(f)\phi\big(f(\mathbf{x}_1, \mathbf{x}_2, \ldots, \mathbf{x}_k)\big) \leq 0. \tag{2.2}$$

If ω is a weighted polymorphism of ϕ, we say ϕ is *improved* by ω.

Note that if ϕ is improved by the weighting $\omega \in \mathbf{W}_C^{(k)}$, then every element of $C^{(k)}$ must be a polymorphism of ϕ. Thus weighted polymorphisms capture both polymorphisms and fractional polymorphisms.

Example 2.8 The submodular multimorphism $\langle \mathrm{Min}, \mathrm{Max} \rangle$ is equivalent to the 2-ary weighted polymorphism ω defined by

$$\omega(f) \stackrel{\text{def}}{=} \begin{cases} -1 & \text{if } f \in \{e_1^{(2)}, e_2^{(2)}\}, \\ 0 & \text{otherwise.} \end{cases}$$

Remark 2.7 A fractional operation, defined in Definition 1.9 (see also Remark 1.17), is a weighting, defined in Definition 2.3, that assigns weight -1 to all projections. We now show that fractional operations and weightings are the same. Consequently, weighted polymorphisms are the same as fractional polymorphisms. Given a k-ary weighting ω, let

$$\mathcal{F} = \big\{\langle \omega(f), f \rangle \big| f \in \mathbf{dom}(\omega) \wedge f \neq e_i^{(k)}, 1 \leq i \leq k\big\}.$$

Let $c_i = |\omega(e_i^{(k)})|$ be the absolute value of the weight given to the ith projection, and define $c_i = 0$ if $e_i^{(k)} \notin \mathbf{dom}(\omega)$. Let $c = \max_{1 \leq i \leq k} c_i$. If $c = 0$, then we define $\mathcal{F}$ to be $\{\langle +1, e_i^{(k)} \rangle\}$. Otherwise, we add projections to $\mathcal{F}$ as follows: we add $\langle c + c_i, e_i^{(k)} \rangle$ to $\mathcal{F}$, if $c + c_i > 0$, for all $1 \leq i \leq k$. Finally, we divide all weights in $\mathcal{F}$ by c.

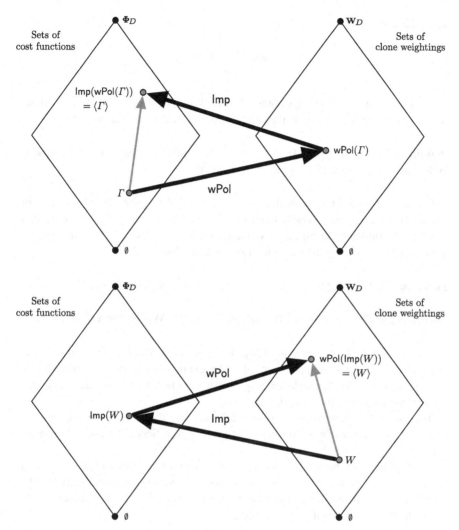

Fig. 2.3 Galois connection between $\mathbf{\Phi}_D$ and $\mathbf{W}_D$

Notation 2.3 We denote by $\mathbf{\Phi}_D$ the set of all cost functions defined on the set D.

Definition 2.8 For any $\Gamma \subseteq \mathbf{\Phi}_D$, we denote by $\mathsf{wPol}(\Gamma)$ the set of all weightings of $\mathsf{Pol}(\Gamma)$ that are weighted polymorphisms of all cost functions $\phi \in \Gamma$.

To define a mapping in the other direction, we need to consider the union of the sets $\mathbf{W}_C$ over all clones C on some fixed set D, which will be denoted by $\mathbf{W}_D$. If we have a set $W \subseteq \mathbf{W}_D$ that may contain weightings of *different* clones over D, then we can extend each of these weightings with zeros, as necessary, so that they are weightings of the same clone C, given by

$$C = Clone(\bigcup_{\omega \in W} \mathbf{dom}(\omega)).$$

This set of extended weightings obtained from W will be denoted by $\overline{W}$. For any set $W \subseteq \mathbf{W}_D$, we define $\langle W \rangle$ to be the smallest weighted clone containing $\overline{W}$.

Definition 2.9 For any $W \subseteq \mathbf{W}_D$, we denote by $\mathsf{Imp}(W)$ the set of all cost functions in $\mathbf{\Phi}_D$ that are improved by all weightings $\omega \in W$.

It follows immediately from the definition of a Galois connection [31] that, for any set D, the mappings $\mathsf{wPol}(\cdot)$ and $\mathsf{Imp}(\cdot)$ form a Galois connection between $\mathbf{W}_D$ and $\mathbf{\Phi}_D$, as illustrated in Fig. 2.3. A characterisation of this Galois connection for finite sets D is given by the following two theorems [68].

Theorem 2.5 *For any finite set D, and any finite $\Gamma \subseteq \mathbf{\Phi}_D$, $\mathsf{Imp}(\mathsf{wPol}(\Gamma)) = \langle \Gamma \rangle$.*

Theorem 2.6 *For any finite set D, and any finite $W \subseteq \mathbf{W}_D$, $\mathsf{wPol}(\mathsf{Imp}(W)) = \langle W \rangle$.*

As with any Galois connection [31], this means that there is a one-to-one correspondence between languages closed under expressibility and weighted clones. Hence, the search for tractable languages over a finite set corresponds to a search for suitable weighted clones of operations.

The proofs of Theorems 2.6 and 2.5 rely on the application of Farkas' Lemma [256] and are based on the ideas presented in Sects. 2.4 and 2.5, respectively.

Creed and Živný have used the algebraic theory to classify so-called minimal Boolean languages [85], thus obtaining (a simpler proof of) the hardness part of the complexity classification of Boolean languages [67] (cf. Chap. 6), where the original results from [67] relied on ad hoc gadgets.

2.7 Summary

We have investigated the expressive power of crisp and valued constraints. We have presented a construction to determine whether a given cost function is expressible over a given language. We have also presented a construction to determine whether a given fractional polymorphism belongs to a fractional clone. We have then presented a general algebraic theory. This algebraic theory allowed for a simpler proof of the (hardness part of the) classification of Boolean languages [85], and also a simpler proof of the (hardness part of the) classification of conservative languages [86], originally obtained in [187, 188] (more in Chap. 7).

2.7.1 Related Work

Galois connections for various variants of the CSP have been considered in the literature [31, 255].

2.7.2 Open Problems

The most interesting open problem is the structure of weighted clones. Due to the presented Galois connection, this would immediately shed some light on the complexity of VCSPs.

Chapter 3
Expressibility of Fixed-Arity Languages

I just wondered how things were put together.
Claude Shannon

3.1 Introduction

In this chapter, we present our results on the expressive power of various classes of valued constraints. Most of the results are of the following form: let $\mathcal{C}$ be a class of valued constraints with cost functions of unbounded arities; then $\mathcal{C}$ can be expressed by a subset of $\mathcal{C}$ consisting of valued constraints with cost functions of a fixed bounded arity. The only known class for which this is not true is the class of finite-valued max-closed cost functions of different arities.

This chapter is organised as follows. In Sect. 3.2, we define various classes of cost functions, and present our results. In Sects 3.3, 3.4, and 3.5, we present our results for *crisp*, *finite-valued*, and *general* cost functions, respectively. We present both algebraic proofs of most results, which have been published in [70], and also several alternative, non-algebraic proofs of some of the results, which have been published in [272]. Finally, in Sect. 3.6, we present some more results on the algebraic properties of finite-valued max-closed cost functions.

3.2 Results

First, we present our results on the expressive power of valued constraint languages containing all cost functions up to some fixed arity over some fixed domain.

Recall that general-valued cost functions can take on both finite and infinite costs.

Definition 3.1 For every $d \geq 2$ we define the following:

Most of the material in this chapter is reprinted from *Theoretical Computer Science*, **409**(1), D.A. Cohen, P.G. Jeavons, and S. Živný, The Expressive Power of Valued Constraints: Hierarchies and Collapses, 137–153, Copyright (2008), with permission from Elsevier and from *Information Processing Letters*, **109**(11), B. Zanuttini and S. Živný, A Note on Some Collapse Results of Valued Constraints, 534–538, Copyright (2009), with permission from Elsevier.

- $\mathbf{R}_{d,m}$ denotes the set of all relations of arity at most m over a domain of size d, and $\mathbf{R}_d \stackrel{\text{def}}{=} \bigcup_{m \geq 0} \mathbf{R}_{d,m}$.
- $\mathbf{F}_{d,m}$ denotes the set of all finite-valued cost functions of arity at most m over a domain of size d, and $\mathbf{F}_d \stackrel{\text{def}}{=} \bigcup_{m \geq 0} \mathbf{F}_{d,m}$.
- $\mathbf{G}_{d,m}$ denotes the set of all general-valued cost functions of arity at most m over a domain of size d, and $\mathbf{G}_d \stackrel{\text{def}}{=} \bigcup_{m \geq 0} \mathbf{G}_{d,m}$.

We will prove the following theorem by showing that crisp cost functions of a fixed arity can express crisp cost functions of arbitrary arities, and that same holds for both finite-valued and general-valued cost functions.

Theorem 3.1 *For all $d \geq 3$ and $f \geq 2$:*

1. $\langle \mathbf{R}_{2,1} \rangle \subsetneq \langle \mathbf{R}_{2,2} \rangle \subsetneq \langle \mathbf{R}_{2,3} \rangle = \mathbf{R}_2$.
2. $\langle \mathbf{R}_{d,1} \rangle \subsetneq \langle \mathbf{R}_{d,2} \rangle = \mathbf{R}_d$.
3. $\langle \mathbf{F}_{f,1} \rangle \subsetneq \langle \mathbf{F}_{f,2} \rangle = \mathbf{F}_f$.
4. $\langle \mathbf{G}_{f,1} \rangle \subsetneq \langle \mathbf{G}_{f,2} \rangle = \mathbf{G}_f$.

We will then consider important subsets of these languages defined for totally-ordered domains, containing the so-called *max-closed* cost functions, which are defined below.

Recall that the function Max denotes the standard binary function which returns the larger of its two arguments.

Definition 3.2 A cost function ϕ is called *max-closed* if $\{\langle 2, \text{Max} \rangle\} \in \text{fPol}(\{\phi\})$.

Observation 3.1 Equivalently, ϕ is max-closed if $\langle \text{Max}, \text{Max} \rangle \in \text{Mul}(\{\phi\})$.

Definition 3.3 For every $d \geq 2$ we define the following:

- $\mathbf{R}_{d,m}^{\max}$ denotes the set of all max-closed relations of arity at most m over an ordered domain of size d, and $\mathbf{R}_d^{\max} \stackrel{\text{def}}{=} \bigcup_{m \geq 0} \mathbf{R}_{d,m}^{\max}$.
- $\mathbf{F}_{d,m}^{\max}$ denotes the set of all finite-valued max-closed cost functions of arity at most m over an ordered domain of size d, and $\mathbf{F}_d^{\max} \stackrel{\text{def}}{=} \bigcup_{m \geq 0} \mathbf{F}_{d,m}^{\max}$.
- $\mathbf{G}_{d,m}^{\max}$ denotes the set of all general-valued max-closed cost functions of arity at most m over an ordered domain of size d, and $\mathbf{G}_d^{\max} \stackrel{\text{def}}{=} \bigcup_{m \geq 0} \mathbf{G}_{d,m}^{\max}$.

We will show below that the following theorem holds for these sets of max-closed cost functions. Note that the result establishes an infinite hierarchy for finite-valued max-closed cost functions.

Theorem 3.2 *For all $d \geq 3$ and $f \geq 2$:*

1. $\langle \mathbf{R}_{2,1}^{\max} \rangle \subsetneq \langle \mathbf{R}_{2,2}^{\max} \rangle \subsetneq \langle \mathbf{R}_{2,3}^{\max} \rangle = \mathbf{R}_2^{\max}$.

2. $\langle \mathbf{R}_{d,1}^{\max} \rangle \subsetneq \langle \mathbf{R}_{d,2}^{\max} \rangle = \mathbf{R}_d^{\max}$.

3. $\langle \mathbf{F}_{f,1}^{\max} \rangle \subsetneq \langle \mathbf{F}_{f,2}^{\max} \rangle \subsetneq \langle \mathbf{F}_{f,3}^{\max} \rangle \subsetneq \langle \mathbf{F}_{f,4}^{\max} \rangle \cdots$

4. $\langle \mathbf{G}_{2,1}^{\max} \rangle \subsetneq \langle \mathbf{G}_{2,2}^{\max} \rangle \subsetneq \langle \mathbf{G}_{2,3}^{\max} \rangle = \mathbf{G}_2^{\max}$.

5. $\langle \mathbf{G}_{d,1}^{\max} \rangle \subsetneq \langle \mathbf{G}_{d,2}^{\max} \rangle = \mathbf{G}_d^{\max}$.

In this chapter, we will prove Theorems 3.1 and 3.2. For some results, we present both an algebraic and a non-algebraic proof. Quite often the algebraic proofs are more involved. However, these proofs provide us with more than a statement of the result; they show us the structure of the algebraic properties of the corresponding class of cost functions. Moreover, for the separating result in Theorem 3.2(3), the algebraic properties play a crucial role. In Sect. 3.5, we characterise the fractional clone of general-valued max-closed cost functions. In Sect. 3.6, we characterise the multimorphisms and fractional polymorphisms of finite-valued max-closed cost functions.

3.3   Crisp Cost Functions

In this section, we consider the expressive power of valued constraint languages containing only *crisp* cost functions, that is, *relations*.

We consider the languages containing all relations up to some fixed arity over some fixed domain, and the languages containing all *max-closed* relations up to some fixed arity over some fixed totally-ordered domain. In both cases, we show that the relations of a fixed arity can express all relations of arbitrary arities.

The class of *crisp* max-closed cost functions was first introduced (as a class of relations) in [167] and shown to be tractable. In other words, VCSP(Γ) is known to be polynomial-time solvable for any set Γ consisting of max-closed relations over any finite set D. A number of examples of max-closed relations are given in [167].

Remark 3.1 The max-closed property generalises the *X-underbar* property studied in the context of graph homomorphisms [150], which applies only to binary relations.

It is well known that any relation can be expressed as a propositional formula in Conjunctive Normal Form (CNF), simply as a conjunction of clauses which disallow tuples not in the relation. Hence we have the following characterisation of $\mathbf{R}_{d,m}$.

Proposition 3.1 *A relation $R \in \mathbf{R}_{d,m}$ if, and only if, there is some formula ψ such that $\langle v_1, \ldots, v_m \rangle \in R \Leftrightarrow \psi(v_1, \ldots, v_m)$ and ψ is a conjunction of clauses of the form $(v_1 \neq a_1) \vee \cdots \vee (v_m \neq a_m)$ for some constants $a_1, \ldots, a_m$.*

We also have a similar characterisation for $\mathbf{R}_{d,m}^{\max}$, adapted from Theorem 5.2 of [167]. (The same result has also been obtained in [129].)

Theorem 3.3 ([167]) *A relation $R \in \mathbf{R}_{d,m}^{\max}$ if, and only if, there is some formula ψ such that $\langle v_1, \ldots, v_m \rangle \in R \Leftrightarrow \psi(v_1, \ldots, v_m)$ and ψ is a conjunction of clauses of the form $(v_1 > a_1) \vee \cdots \vee (v_m > a_m) \vee (v_i < b_i)$ for some constants $a_1, \ldots, a_m, b_i$.*

Note that in the special case of a Boolean domain (that is, when $d = 2$) this restricted form of clauses is equivalent to a disjunction of literals with at most one negated literal; clauses of this form are sometimes called *anti-Horn* clauses.

It is well known that for every $d \geq 2$, $\mathsf{Pol}(\mathbf{R}_d)$ is equal to the set of all possible projection operations [101]. We now characterise the polymorphisms of $\mathbf{R}_d^{\max}$.

Definition 3.4 Let D be a fixed totally-ordered set.

- The k-ary function on D that returns the largest of its k arguments in the given ordering of D is denoted by Max_k.
- The k-ary function on D that returns the smallest of its k arguments in the given ordering of D is denoted by Min_k.
- The k-ary function on D that returns the second largest of its $k \geq 2$ arguments in the given ordering of D is denoted by SECOND_k.

The function Max_2 will be denoted by Max and the function Min_2 will be denoted by Min.

Definition 3.5 Let $I = \{i_1, \ldots, i_n\} \subseteq \{1, \ldots, k\}$ be a set of indices. Define the k-ary function

$$\mathrm{Max}_I(x_1, \ldots, x_k) \stackrel{\mathrm{def}}{=} \mathrm{Max}_n(x_{i_1}, \ldots, x_{i_n}).$$

For every k, there are exactly $2^k - 1$ functions of the form Max_I for $\emptyset \neq I \subseteq \{1, \ldots, k\}$.

Proposition 3.2 *For all $d \geq 2$,*

$$\mathsf{Pol}\left(\mathbf{R}_d^{\max}\right) = \left\{\mathrm{Max}_I \mid \emptyset \neq I \subseteq \{1, \ldots, k\}, k = 1, 2, \ldots\right\}.$$

Proof When $|I| = 1$, the corresponding function Max_I is just a projection operation, and every projection is a polymorphism of every relation (cf. Observation 1.1).

If $\mathrm{Max} \in \mathsf{Pol}(\{R\})$, then $\mathrm{Max}_I \in \mathsf{Pol}(\{R\})$ for every $\emptyset \neq I \subseteq \{1, \ldots, k\}$. This is because $\mathsf{Pol}(\{R\})$ is closed under function composition and contains all projection operations, and every Max_I can be obtained by function composition from the function Max and the projection operations.

We now prove that the operations of the form Max_I are the *only* polymorphisms of $\mathbf{R}_d^{\max}$. Suppose, for contradiction, that f is a k-ary polymorphism of $\mathbf{R}_d^{\max}$ different from Max_I for every $\emptyset \neq I \subseteq \{1, \ldots, k\}$. It follows that, for each I such that $\emptyset \neq I \subseteq \{1, \ldots, k\}$, there is a k-tuple t_I, such that $f(t_I) \neq \mathrm{Max}_I(t_I)$. Let n be the total number of different tuples t_I, that is, $n = |\{t_I \mid \emptyset \neq I \subseteq \{1, \ldots, k\}\}| \leq 2^k - 1$, and denote these tuples by $t_1, \ldots, t_n$. Now consider the n-ary relation $R =$

$\{\langle t_1[j], \ldots, t_n[j] \rangle\}_{1 \le j \le k}$. Define $R_0 = R$ and $R_{i+1} = R_i \cup \{\text{Max}(u, v) \mid u, v \in R_i\}$ for every $i \ge 0$. Clearly, $R_i \subseteq R_{i+1}$ and since there is only a finite number of different n-tuples, there is an l such that $R_l = R_{l+i}$ for every $i \ge 0$. Define R' to be the closure of R under Max, that is, $R' = R_l$. Clearly, R' is max-closed and every tuple t of R' is of the form $t = \text{Max}_j(u_{i_1}, \ldots, u_{i_j})$ for some $j \ge 1$ and $u_{i_1}, \ldots, u_{i_j} \in R$. We have constructed R so that the application of f to the tuples of R results in a tuple t that is different from every tuple of this form, and hence $t \notin R'$. Therefore, $f \notin \text{Pol}(\{R'\})$, which means that $f \notin \text{Pol}(\mathbf{R}_d^{\text{max}})$.   □

We now consider the expressive power of $\mathbf{R}_{d,m}$ and $\mathbf{R}_{d,m}^{\text{max}}$. It is clear that binary relations have greater expressive power than unary relations, so our first result is not unexpected, but it provides a simple illustration of the use of the algebraic approach.

Proposition 3.3 *For all* $d \ge 2$, $\langle \mathbf{R}_{d,1} \rangle \subsetneq \langle \mathbf{R}_{d,2} \rangle$ *and* $\langle \mathbf{R}_{d,1}^{\text{max}} \rangle \subsetneq \langle \mathbf{R}_{d,2}^{\text{max}} \rangle$.

Proof Notice for example that Min $\in \text{Pol}(\mathbf{R}_{d,1})$ and consequently Min $\in \text{Pol}(\mathbf{R}_{d,1}^{\text{max}})$ but Min $\notin \text{Pol}(\mathbf{R}_{d,2})$ and Min $\notin \text{Pol}(\mathbf{R}_{d,2}^{\text{max}})$. The result then follows from Theorem 1.2.   □

As a first step, we now focus on the special case of relations over a Boolean domain, that is, the case when $d = 2$. This special case has been studied in detail in [30]. Here, we give a brief independent derivation of the relevant results using the techniques introduced above. We first show that the set of all ternary relations over a Boolean domain has fewer polymorphisms than the set of all binary relations, and hence has greater expressive power. We also establish similar results for max-closed relations over a Boolean domain.

Proposition 3.4 MAJORITY $\in \text{Pol}(\mathbf{R}_{2,2})$ *and* MAJORITY $\in \text{Pol}(\mathbf{R}_{2,2}^{\text{max}})$, *where* MAJORITY *is the unique ternary function on a two-element set which returns the argument value that occurs most often.*

Proof Let R be an arbitrary binary Boolean relation. Let $a = \langle a_1, a_2 \rangle$, $b = \langle b_1, b_2 \rangle$, and $c = \langle c_1, c_2 \rangle$ be three pairs belonging to R. Note that since the domain size is 2, the pair $\langle \text{MAJORITY}(a_1, b_1, c_1), \text{MAJORITY}(a_2, b_2, c_2) \rangle$ is equal to at least one of a, b, c, and hence belongs to R.   □

Proposition 3.5 MAJORITY $\notin \text{Pol}(\mathbf{R}_{2,3})$ *and* MAJORITY $\notin \text{Pol}(\mathbf{R}_{2,3}^{\text{max}})$.

Proof Consider the ternary Boolean max-closed relation R consisting of all triples except $\langle 0, 0, 0 \rangle$. To see that MAJORITY is not a polymorphism of R, consider the triples $\langle 0, 0, 1 \rangle$, $\langle 0, 1, 0 \rangle$, and $\langle 1, 0, 0 \rangle$. The application of MAJORITY to these tuples results in the triple $\langle 0, 0, 0 \rangle$, which is not in R.   □

However, we now show that *ternary* Boolean relations have the same expressive power as *all* Boolean relations. In other words, any Boolean relation of arbitrary arity is expressible by relations of arity at most 3. The same result also holds for max-closed Boolean relations.

Proposition 3.6 $\mathbf{R}_2 \subseteq \langle \mathbf{R}_{2,3} \rangle$ *and* $\mathbf{R}_2^{max} \subseteq \langle \mathbf{R}_{2,3}^{max} \rangle$.

Proof By Proposition 3.1, any Boolean relation $R \in \mathbf{R}_2$ can be expressed as a CNF formula ψ. By the standard SATISFIABILITY to 3-SAT reduction [125], there is a 3-CNF formula ψ' expressing R such that ψ is satisfiable if, and only if, ψ' is satisfiable.

Since the standard SATISFIABILITY to 3-SAT reduction preserves the anti-Horn form of clauses, the same result holds for max-closed Boolean relations.   □

Combining these results with Theorem 1.2, we obtain the following result.

Theorem 3.4

1. $\langle \mathbf{R}_{2,1} \rangle \subsetneq \langle \mathbf{R}_{2,2} \rangle \subsetneq \langle \mathbf{R}_{2,3} \rangle = \mathbf{R}_2$.
2. $\langle \mathbf{R}_{2,1}^{max} \rangle \subsetneq \langle \mathbf{R}_{2,2}^{max} \rangle \subsetneq \langle \mathbf{R}_{2,3}^{max} \rangle = \mathbf{R}_2^{max}$.

For relations over a domain with three or more elements, similar results can be obtained. In fact, in this case we show that *any* relation can be expressed using *binary* relations.

Proposition 3.7 *For all* $d \geq 3$, $\mathbf{R}_d \subseteq \langle \mathbf{R}_{d,2} \rangle$.

Proof Without loss of generality, assume that $D = \{0, \ldots, M\}$, where $M = d - 1$. Define the binary relation R_d by

$$R_d \stackrel{\text{def}}{=} \{\langle 0, i \rangle, \langle i, 0 \rangle \mid 0 \leq i \leq M\} \cup \{\langle i, i+1 \rangle \mid 1 \leq i < M\}.$$

It is known that the only polymorphisms of the relation R_d are projection operations [108]. Hence, by Theorem 1.2, $\langle \{R_d\} \rangle = \mathbf{R}_d$.   □

We now present a non-algebraic proof.

Proof (Alternative proof of Proposition 3.7) By Proposition 3.1, every relation is logically equivalent to some conjunction of clauses. Therefore, Proposition 3.6 gives a weaker result, namely that $\mathbf{R}_d \subseteq \langle \mathbf{R}_{d,3} \rangle$. We need to show how to express a clause C of length 3 (over the domain D, where $d = |D| \geq 3$) as a conjunction of clauses of length 2. Let $D = \{1, \ldots, d\}$ and $C = (U_1(x_1) \vee U_2(x_2) \vee U_3(x_3))$ for some literals (unary relations) U_i, $1 \leq i \leq 3$. We claim that C is equivalent to $\exists y C' = (U_1(x_1) \vee N_1(y)) \wedge (U_2(x_2) \vee N_2(y)) \wedge (U_3(x_3) \vee N_3(y))$, where y is a new variable and $N_1(y) = D \setminus \{1\}$ ("not 1"), $N_2(y) = D \setminus \{2\}$, and $N_3(y) = \{1, 2\}$. It is not difficult to see that a satisfying assignment of C can be extended to a satisfying assignment of C', and conversely, a satisfying assignment of C' gives a satisfying assignment of C.   □

By investigating the polymorphisms of binary max-closed relations, we now show that max-closed relations over non-Boolean domains can also be expressed using binary relations.

Theorem 3.5 *For all* $d \geq 3$, $\mathbf{R}_d^{\max} \subseteq \langle \mathbf{R}_{d,2}^{\max} \rangle$.

Proof We will show that $\mathsf{Pol}(\mathbf{R}_{d,2}^{\max}) \subseteq \mathsf{Pol}(\mathbf{R}_d^{\max})$. The result then follows from Theorem 1.2.

Without loss of generality, assume that $D = \{0, \ldots, M\}$, where $M = d - 1$. Let $f \in \mathsf{Pol}(\mathbf{R}_{d,2}^{\max})$ be an arbitrary k-ary polymorphism. By Proposition 3.2, it is enough to show that $f = \mathrm{Max}_I$ for some $\emptyset \neq I \subseteq \{1, \ldots, k\}$.

First note that for any subset $S \subseteq D$, the binary relation $R = \{\langle a, a \rangle \mid a \in S\}$ is max-closed, so $f(x_1, \ldots, x_k) \in \{x_1, \ldots, x_k\}$. In other words, f is conservative.

If $f = \mathrm{Max}_{\{1,\ldots,k\}}$ we are done. Otherwise, there exist $a_1, \ldots, a_k \in D$ such that $a_i = \mathrm{Max}_k(a_1, \ldots, a_k)$ and $a_i > f(a_1, a_2, \ldots, a_k) = a_j$. Without loss of generality, in order to simplify our notation, assume that $i = 1$ and $j = 2$, that is, $a_1 = \mathrm{Max}_k(a_1, \ldots, a_k)$ and $a_1 > f(a_1, a_2, \ldots, a_k) = a_2$. We will show that f does not depend on its first parameter.

For any fixed $x_2, \ldots, x_k \in D$, we denote the tuple $\langle x_2, \ldots, x_k \rangle$ by $\bar{x}$, and we define the binary max-closed relation

$$R_{\bar{x}} \stackrel{\text{def}}{=} \left(\{a_2, \ldots, a_k\} \times \{x_2, \ldots, x_k\}\right) \cup \left(\{a_1\} \times D\right).$$

Now consider the function $g_{\bar{x}}(r) = f(r, x_2, \ldots, x_k)$. Note that $g_{\bar{x}}(r)$ is a restriction of f with all arguments except the first one fixed.

Claim 1 $\forall r \in D$, $g_{\bar{x}}(r) \in \{x_2, \ldots, x_k\}$.

To establish this claim, note that for all $r \in D$ we have $\langle a_1, r \rangle \in R_{\bar{x}}$, and $\{\langle a_j, x_j \rangle \mid j = 2, \ldots, k\} \subseteq R_{\bar{x}}$. Since f is a polymorphism of $R_{\bar{x}}$ and $f(a_1, a_2, \ldots, a_k) = a_2$, it follows from the definition of $R_{\bar{x}}$ that $g_{\bar{x}}(r) \in \{x_2, \ldots, x_k\}$.

Now we show that if the largest element of the domain, M, is not among $x_2, \ldots, x_k$, then $g_{\bar{x}}(r)$ is constant.

Claim 2 $M \notin \{x_2, \ldots, x_k\} \Rightarrow \forall r \in D$, $g_{\bar{x}}(r) = g_{\bar{x}}(M)$.

To establish this claim, define the binary max-closed relation

$$R_{\bar{x}}' \stackrel{\text{def}}{=} \left(\{M\} \times D\right) \cup \left\{\langle x_j, x_j \rangle \mid j = 2, \ldots, k\right\}.$$

For all $r \in D$ we have $\langle M, r \rangle \in R_{\bar{x}}'$ and $\{\langle x_j, x_j \rangle \mid j = 2, \ldots, k\} \subseteq R_{\bar{x}}'$. By Claim 1, $g_{\bar{x}}(M) = x_i$ for some $2 \leq i \leq k$. Since f is a polymorphism of $R_{\bar{x}}'$, it follows from the definition of $R_{\bar{x}}'$ that $g_{\bar{x}}(r) = x_i = g_{\bar{x}}(M)$ for every $r \in D$.

Next we generalise Claim 2 to show that $g_{\bar{x}}(r)$ is constant whenever $x_2, \ldots, x_k$ does not contain all elements of the domain D.

Claim 3 $\{x_2, \ldots, x_k\} \neq D \Rightarrow \forall r \in D, \ g_{\bar{x}}(r) = g_{\bar{x}}(M)$.

To establish this claim, we will show that for every $p \in D$, if $p \notin \{x_2, \ldots, x_k\}$, then $g_{\bar{x}}(r) = g_{\bar{x}}(M)$ for every $r \in D$. Note that the case $p = M$ is already proved by Claim 2. For any $p \in D \setminus \{M\}$, define the binary max-closed relation $R_p = \{\langle d, \Delta_p(d) \rangle \mid d \in D\}$, where

$$\Delta_p(x) \overset{\text{def}}{=} \begin{cases} x & \text{if } x \leq p, \\ x - 1 & \text{if } x > p. \end{cases}$$

For all $r \in D$ we have $\langle r, \Delta_p(r) \rangle \in R_p$ and $\{\langle x_j, \Delta_p(x_j) \rangle \mid j = 2, \ldots, k\} \subseteq R_p$. Since f is a polymorphism of R_p, it follows from the definition of R_p that for every $r \in D$, $g_{\bar{x}}(r) \in \Delta_p^{-1}(g_{\Delta_p(\bar{x})}(\Delta_p(r)))$.

Since $M \notin \{\Delta_p(d) \mid d \in D\}$, we know, by Claim 2, that $g_{\Delta_p(\bar{x})}(\Delta_p(r))$ is constant. Say $g_{\Delta_p(\bar{x})}(\Delta_p(r)) = k_p$. If $k_p \neq p$, then $|\Delta_p^{-1}(k_p)| = 1$ and so $g_{\bar{x}}$ is constant. Alternatively, if $k_p = p$, then $\Delta_p^{-1}(k_p) = \{p, p+1\}$. In this case if $p \notin \{x_2, \ldots, x_k\}$, then we know, by Claim 1, that $g_{\bar{x}}(r) \neq p$, so $g_{\bar{x}}$ is again constant. This completes the proof of Claim 3.

Claim 4 $g_{\bar{x}}(r)$ *is constant*.

To establish this claim, define the binary max-closed relations $R_+ = \{\langle d, \Delta_+(d) \rangle \mid d \in D\}$ and $R_- = \{\langle d, \Delta_-(d) \rangle \mid d \in D\}$, where

$$\Delta_+(x) \overset{\text{def}}{=} \begin{cases} x & \text{if } x \neq M, \\ x - 1 & \text{if } x = M, \end{cases}$$

and

$$\Delta_-(x) \overset{\text{def}}{=} \begin{cases} x & \text{if } x \neq 0, \\ x + 1 & \text{if } x = 0. \end{cases}$$

Define $\bar{y} = \langle \Delta_+(x_2), \ldots, \Delta_+(x_k) \rangle$ and $\bar{z} = \langle \Delta_-(x_2), \ldots, \Delta_-(x_k) \rangle$. Since $M \notin \{\Delta_+(d) \mid d \in D\}$ and $0 \notin \{\Delta_-(d) \mid d \in D\}$, we know, by Claim 3, that $g_{\bar{y}}$ and $g_{\bar{z}}$ are both constant.

For every $r \in D$, $\langle r, \Delta_+(r) \rangle \in R_+$ and for every $i = 2, \ldots, k$, $\langle x_i, \Delta_+(x_i) \rangle \in R_+$. Since f is a polymorphism of R_+, and $g_{\bar{y}}$ is constant, $g_{\bar{x}}$ is either constant or for every $r \in D$, $g_{\bar{x}}(r) \in \{M, M - 1\}$. Similarly, for every $r \in D$, $\langle r, \Delta_-(r) \rangle \in R_-$ and for every $i = 2, \ldots, k$, $\langle x_i, \Delta_-(x_i) \rangle \in R_-$. Since f is a polymorphism of R_-, and $g_{\bar{z}}$ is constant, $g_{\bar{x}}$ is either constant or for every $r \in D$, $g_{\bar{x}}(r) \in \{0, 1\}$. Since $|D| > 2$ we know[1] that $|\{M, M - 1\} \cap \{0, 1\}| \leq 1$. Hence, in all cases $g_{\bar{x}}$ is constant.

We have shown that if $a_1 = \max(a_1, \ldots, a_k)$ and $f(a_1, \ldots, a_k) < a_1$, then f does not depend on its first parameter. Similarly, by repeating the same argument, we can show that if $f \neq \text{Max}_{\{2, \ldots, k\}}$, then f does not depend on its ith parameter for some

---

[1] This is the only place where we use the condition that $|D| \geq 3$.

i such that $2 \leq i \leq k$. Moreover, further repeating the same argument shows that if f does not depend on any parameter outside of $I \subseteq \{1, \ldots, k\}$ and $f \neq \mathrm{Max}_I$, then f does not depend on any of the parameters whose index is in I.

Therefore, either there is some set $I \subseteq \{1, \ldots, k\}$ for which $f = \mathrm{MAX}_I$ or f is constant. However, since f is conservative, it cannot be constant. □

Next we present a non-algebraic proof.

Proof (Alternative proof of Theorem 3.5) It is enough to show that any clause of the form $(x_1 \geq a_1 \vee \cdots \vee x_k \geq a_k)$ or $(x_1 \geq a_1 \vee \cdots \vee x_k \geq a_k \vee y \leq b)$, where $a_1, \ldots, a_k, b$ are domain values, can be expressed by a conjunction of anti-Horn clauses over at most two variables.

Let C be a clause in $\mathbf{R}_d^{\max}$:

$$C = (x_1 \geq a_1 \vee x_2 \geq a_2 \vee \cdots \vee x_k \geq a_k \vee y \leq b)$$

(the case without the y literal is even easier and can be handled similarly).

For all $i = 1, \ldots, k - 1$, let y_i be a fresh variable. As $d \geq 3$, the domain of each y_i contains at least three different values, say $1, 2,$ and 3 with the natural order. We define the following conjunction of clauses ψ, where $y_i \in \{1, 3\}$ is used as a shorthand for $y_i \geq 3 \vee y_i \leq 1$ (possible values less than 1 or greater than 3 do not matter) as follows:

$$\psi \overset{\mathrm{def}}{=} \left(x_1 \geq a_1 \vee y_1 \in \{1, 3\} \right) \wedge (y_1 \leq 2 \vee x_2 \geq a_2)$$

$$\wedge \bigwedge_{i=2}^{k-1} \left(\begin{array}{c} (y_{i-1} \geq 2 \vee y_i \in \{1, 3\}) \\ \wedge (y_i \leq 2 \vee x_{i+1} \geq a_{i+1}) \end{array} \right)$$

$$\wedge (y_{k-1} \geq 2 \vee y \leq b).$$

The intuition is given by reading the second clause as $(y_1 \geq 3 \rightarrow x_2 \geq a_2)$ and the third one as $(y_1 \leq 1 \rightarrow y_2 \in \{1, 3\})$. Since the first clause reads "$x_1 \geq a_1$ or $y_1 \geq 3$ or $y_1 \leq 1$", together with the above implications this gives "$x_1 \geq a_1$ or $x_2 \geq a_2$ or $y_2 \in \{1, 3\}$". Iterating this reasoning, one can see intuitively why the construction works.

More formally, we show that C is logically equivalent to $\exists y_1 \ldots y_{k-1} \psi$. First, let t be a tuple satisfying ψ. Then if t satisfies $x_1 \geq a_1$, we are done. Otherwise, because of the first clause in ψ, t must satisfy (1) $y_1 \geq 3$ or (2) $y_1 \leq 1$. In case (1), because of the second clause in ψ, t must satisfy $x_2 \geq a_2$ and we are done. In case (2), because of the third clause in ψ, t must satisfy $y_2 \geq 3 \vee y_2 \leq 1$, and we proceed by induction.

Conversely, let t be a tuple satisfying C. We show that t can be completed into a model of ψ by assignments to the y_i's.

Assume first that t satisfies $x_1 \geq a_1$. Then completing t with $t(y_i) = 2$ for all $i = 1, \ldots, k - 1$ yields a model of ψ whatever the values assigned by t to $x_2, \ldots, x_k, y$. This can be seen by examining each clause in ψ. Now assume that t satisfies $x_{i_0} \geq a_{i_0}$ for some $i_0 \in \{2, \ldots, k\}$. Then completing t with $t(y_i) = 1$ for

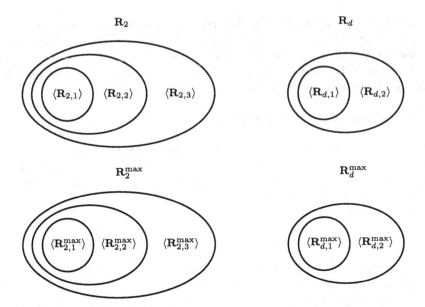

Fig. 3.1 Summary of results from Sect. 3.3, for all $d \geq 3$

all $i = 1, \ldots, i_0 - 2$, $t(y_{i_0 - 1}) = 3$, and $t(y_i) = 2$ for all $i = i_0, \ldots, k - 1$ again
yields a model of ψ. Finally, assume that t satisfies $y \leq b$. Then completing t with
$t(y_i) = 1$ for all $i = 1, \ldots, k - 1$ yields a model of ψ, which finishes the proof.

Note that this proof makes clear why the same argument does not work for $d = 2$.
Indeed, the "hole" in literal $y_i \in \{1, 3\}$ is necessary, since otherwise this literal would
be tautologous and thus every second clause would be in ψ. And Theorem 3.4 indeed shows that in the Boolean case, ternary relations are both sufficient and necessary.   □

Combining these results we obtain the following result:

Theorem 3.6 *For all $d \geq 3$,*

1. $\langle \mathbf{R}_{d,1} \rangle \subsetneq \langle \mathbf{R}_{d,2} \rangle = \mathbf{R}_d$.
2. $\langle \mathbf{R}_{d,1}^{\mathrm{max}} \rangle \subsetneq \langle \mathbf{R}_{d,2}^{\mathrm{max}} \rangle = \mathbf{R}_d^{\mathrm{max}}$.

Figure 3.1 summarises the results from this section.

3.4 Finite-Valued Cost Functions

In this section, we consider the expressive power of valued constraint languages
containing only *finite-valued* cost functions.

Cost functions from $\mathbf{F}_2$, that is, finite-valued cost functions over a Boolean domain, are also known as *pseudo-Boolean functions* [34, 84]. The class of max-closed cost functions is discussed in more detail in [67] and shown to be tractable. A number of examples of max-closed cost functions are given in [67].

First we show that the set of all finite-valued cost functions of a certain fixed arity can express all finite-valued cost functions of arbitrary arities. On the other hand, we show that the *max-closed* finite-valued cost functions of any fixed arity *cannot* express all finite-valued max-closed cost functions of any larger arity. Hence we identify an infinite hierarchy of finite-valued cost functions with ever-increasing expressive power.

Proposition 3.8 *For all $d \geq 2$, $\langle \mathbf{F}_{d,1} \rangle \subsetneq \langle \mathbf{F}_{d,2} \rangle$ and $\langle \mathbf{F}_{d,1}^{\max} \rangle \subsetneq \langle \mathbf{F}_{d,2}^{\max} \rangle$.*

Proof Consider the binary fractional operation $\mathcal{F} = \{\langle 1, \text{Min} \rangle, \langle 1, \text{Max} \rangle\}$. It is straightforward to verify that $\mathcal{F} \in \mathsf{fPol}(\mathbf{F}_{d,1})$ and $\mathcal{F} \in \mathsf{fPol}(\mathbf{F}_{d,1}^{\max})$.

Now consider the binary finite-valued max-closed cost function ϕ over any domain containing $\{0, 1\}$, defined by $\phi(\langle 0, 0 \rangle) = 1$ and $\phi(\langle \cdot, \cdot \rangle) = 0$ otherwise. Note that ϕ is max-closed but $\mathcal{F}$ is *not* a fractional polymorphism of ϕ. To see this, consider the tuples $\langle 0, 1 \rangle$ and $\langle 1, 0 \rangle$ (see the tableau below).

$$\left.\begin{array}{c} 0\ 1 \\ 1\ 0 \end{array}\ \xrightarrow{\phi}\ \begin{array}{c} 0 \\ 0 \end{array}\right\} \Sigma = 0$$

$$\begin{array}{c} \text{Min} \\ \text{Max} \end{array} \left.\begin{array}{c} \overline{0\ 0} \\ 1\ 1 \end{array}\ \xrightarrow{\phi}\ \begin{array}{c} 1 \\ 0 \end{array}\right\} \Sigma = 1$$

The result then follows from Theorem 1.3. □

Now we prove a collapse result for the set of all finite-valued cost functions over an arbitrary finite domain. This result was previously known for the special case when $d = 2$: as we remarked earlier, any Boolean finite-valued cost function can be represented as a pseudo-Boolean function; using a well-known result from pseudo-Boolean optimisation [34, 84], any such function can be expressed using binary pseudo-Boolean functions.

Theorem 3.7 *For all $d \geq 2$, $\langle \mathbf{F}_{d,2} \rangle = \mathbf{F}_d$.*

Proof As mentioned above, the case $d = 2$ follows from well-known results about pseudo-Boolean functions (cf. Theorem 1 of [34]). Let $\phi \in \mathbf{F}_{d,m}$ for some $d \geq 3$ and $m > 2$. We will show how to express ϕ using only unary and binary finite-valued cost functions. Without loss of generality, assume that all cost functions are defined over the set $D = \{0, 1, \dots, M\}$, where $M = d - 1$, and denote by $D^m = \{t_1, \dots, t_n\}$ the set of all m-tuples over D. Clearly, $n = d^m$. Let $K \in \mathbb{Q}_{\geq 0}$ be a fixed constant such that $K > \max_{t \in D^m} \phi(t)$. For any $e \in D$, let χ^e be the binary finite-valued cost

Fig. 3.2 A part of the gadget for expressing ϕ (Theorem 3.7)

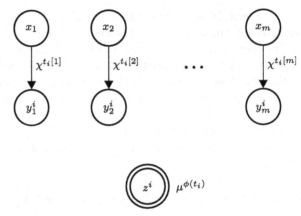

function defined by

$$\chi^e(x, y) \overset{\text{def}}{=} \begin{cases} 0 & \text{if } x = e \text{ and } y = 0, \\ 0 & \text{if } x \neq e \text{ and } y = 1, \\ K & \text{otherwise.} \end{cases}$$

For any $r \in \mathbb{Q}_{\geq 0}$, let μ^r be the unary finite-valued cost function defined by

$$\mu^r(z) \overset{\text{def}}{=} \begin{cases} r & \text{if } z = 0, \\ 0 & \text{otherwise.} \end{cases}$$

We now start building the gadget for ϕ. Let $x_1, \ldots, x_m$ be the variables upon which we wish to construct ϕ, and let $t_i \in D^m$ be an arbitrary fixed tuple. Figure 3.2 shows the part of the gadget for ϕ which ensures that the appropriate cost value is assigned to the tuple of values t_i. The complete gadget for ϕ consists of this part in n copies: one copy with a new set of variables for every $t_i \in D^m$.

Define new variables $y_1^i, \ldots, y_m^i$ and z^i. We apply cost functions on these variables as shown in Fig. 3.2. Note that each variable y_j^i, $1 \leq j \leq m$, indicates whether or not x_j is equal to $t_i[j]$: in any minimum-cost assignment, $(y_j^i = 0) \Leftrightarrow (x_j = t_i[j])$. It remains to define the constraints between the variables $y_1^i, \ldots, y_m^i$, and z^i. These will be chosen in such a way that any assignment of the values 0 or 1 to the variables y_j^i can be extended to an assignment to z^i with a total cost equal to the same fixed minimum value. Furthermore, in these extended assignments z^i is assigned 0 if, and only if, all the y_j^i are assigned 0. (We will achieve this by combining appropriate binary finite-valued cost functions over these variables and other fresh variables as described below.) Then, for every possible assignment of values t_i to the variables $x_1, \ldots, x_m$, there is exactly one z^i, $1 \leq i \leq n$, that is assigned the value 0 in any minimum-cost extension of this assignment. The unary constraint with cost function $\mu^{\phi(t_i)}$ on each z^i then ensures that the complete gadget expresses ϕ.

Fig. 3.3 $\mathcal{P}$ expressing or_2
over non-Boolean domains
(Theorem 3.7)

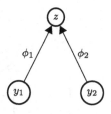

To define the remaining constraints to complete the constraint in Fig. 3.2, we define two binary finite-valued cost functions as follows:

$$\phi_1(y, z) \overset{\text{def}}{=} \begin{cases} 0 & \text{if } y = 0 \text{ and } (z = 0 \text{ or } z = 1), \\ 0 & \text{if } y \neq 0 \text{ and } z \neq 0, \\ K & \text{otherwise,} \end{cases}$$

and

$$\phi_2(y, z) \overset{\text{def}}{=} \begin{cases} 0 & \text{if } y = 0 \text{ and } (z = 0 \text{ or } z = 2), \\ 0 & \text{if } y \neq 0 \text{ and } z \neq 0, \\ K & \text{otherwise.} \end{cases}$$

Let $\mathcal{P} = \langle V, D, C \rangle$, where $V = \{y_1, y_2, z\}$ and $C = \{\langle\langle y_1, z \rangle, \phi_1\rangle, \langle\langle y_2, z \rangle, \phi_2\rangle\}$. (See Fig. 3.3.)

We define or_2 to be the ternary cost function expressed by the gadget $\langle \mathcal{P}, \langle y_1, y_2, z \rangle\rangle$. The cost function $or_2(y_1, y_2, z)$ has the following properties:

- if both y_1, y_2 are assigned the value 0, then the total cost is 0 if, and only if, z is assigned the value 0; otherwise the total cost is either K (if $z = 1$ or $z = 2$) or $2K$ (if $z > 2$);
- if y_1 is assigned the value 0 and y_2 a nonzero value, then the total cost is 0 if, and only if, z is assigned 1; otherwise the total cost is K;
- if y_1 is assigned a nonzero value and y_2 the value 0, then the total cost is 0 if, and only if, z is assigned 2; otherwise the total cost is K;
- if both y_1 and y_2 are assigned nonzero values, then the total cost is 0 if, and only if, z is assigned a nonzero value; otherwise the total cost is $2K$.

All these properties of or_2 can be easily verified by examining the so-called *microstructure* [169] of $\mathcal{P}$, as shown in Fig. 3.4: this is a graph where the vertices are pairs $\langle v, e \rangle \in V \times D$, and two vertices $\langle v_1, e_1 \rangle$ and $\langle v_2, e_2 \rangle$ are connected by an edge with weight w if, and only if, there is a valued constraint $\langle\langle v_1, v_2 \rangle, c \rangle \in C$ such that $c(e_1, e_2) = w$. Circles represent particular assignments to particular variables, as indicated in Fig. 3.4, and edges are weighted by the cost of the corresponding pairs of assignments. Thin edges indicate zero weight, and bold edges indicate weight K.

We have shown that, in any minimum-cost assignment for $\mathcal{P}$, the variable z takes the value 0 if, and only if, both of the variables y_1 and y_2 take the value 0. Hence the cost function or_2 can be viewed as a kind of 2-input "or-gate", with inputs y_1 and y_2 and output z. By cascading $m - 1$ copies of this gadget we can express a cost function $or_m(y_1, \ldots, y_m, z)$, with the following properties:

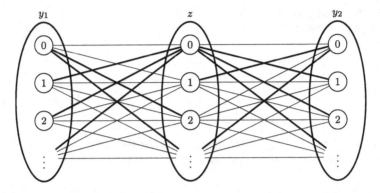

Fig. 3.4 Microstructure of the instance $\mathcal{P}$ (Theorem 3.7)

- if the arguments $y_1, y_2, \ldots, y_m$ are all assigned 0, then assigning zero to z gives cost 0, but any nonzero assignment to z gives cost at least K;
- if not all the arguments $y_1, y_2, \ldots, y_m$ are assigned 0, then there is a nonzero value $e \in D$ such that assigning e to z gives cost 0, but assigning zero to z gives cost at least K.

Using this combined gadget on the variables $y_1^i, y_2^i, \ldots, y_m^i$, and z^i in Fig. 3.2 completes the gadget for ϕ, and hence establishes that $\phi \in \langle \mathbf{F}_{d,2} \rangle$.   □

In contrast to this result, the remaining results in this section establish an infinite hierarchy of increasing expressive power for finite-valued *max-closed* cost functions.

Notation 3.1 We will say that an m-tuple u *dominates* an m-tuple v, denoted by $u \geq v$, if $u[i] \geq v[i]$ for all $1 \leq i \leq m$.

Proposition 3.9 ([67]) *An m-ary cost function $\phi : D^m \to \overline{\mathbb{Q}}_{\geq 0}$ is max-closed if, and only if, $\mathrm{Max} \in \mathrm{Pol}(\{\phi\})$ and ϕ is finitely antitone, that is, for all m-tuples u, v with $\phi(u), \phi(v) < \infty, u \leq v \Rightarrow \phi(u) \geq \phi(v)$.*

It follows that the finite-valued max-closed cost functions are simply the finite-valued antitone functions, that is, those functions whose values can only decrease as their arguments get larger. Note that for such functions the expressive power is likely to be rather limited because in any construction the "hidden variables" that are "projected out" can always be assigned the highest values in their domain in order to minimise the cost. Hence, using such hidden variables only adds a constant value to the total cost, and so does not allow more cost functions to be expressed.

We now extend the separation result shown in Proposition 3.8 and separate each possible arity.

Proposition 3.10 *For all $d \geq 2$ and $m \geq 2$, $\{\langle m - 1, \mathrm{Max}_m \rangle, \langle 1, \mathrm{SECOND}_m \rangle\} \in \mathrm{fPol}(\mathbf{F}_{d,m-1}^{\max})$.*

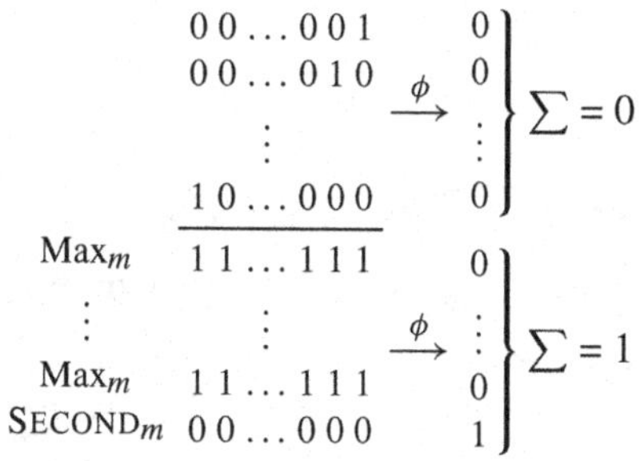

Fig. 3.5
$\{\langle m-1, \text{Max}_m\rangle, \langle 1, \text{Second}_m\rangle\} \notin \text{fPol}(\{\phi\})$
for ϕ (Proposition 3.11)

$$
\left.
\begin{array}{l}
0\,0\dots0\,0\,1 \\
0\,0\dots0\,1\,0 \\
\quad\vdots \\
1\,0\dots0\,0\,0
\end{array}
\xrightarrow{\phi}
\begin{array}{c}
0 \\
0 \\
\vdots \\
0
\end{array}
\right\}\sum = 0
$$

$$
\begin{array}{l l}
\text{Max}_m & \overline{1\,1\dots1\,1\,1} \\
& \quad\vdots \\
\text{Max}_m & 1\,1\dots1\,1\,1 \\
\text{Second}_m & 0\,0\dots0\,0\,0
\end{array}
\xrightarrow{\phi}
\left.
\begin{array}{c}
0 \\
\vdots \\
0 \\
1
\end{array}
\right\}\sum = 1
$$

Proof Let ϕ be an arbitrary $(m-1)$-ary finite-valued max-closed cost function. Let $t_1, \dots, t_m$ be $(m-1)$-tuples. We show that there is an i such that the tuple $s = \text{Second}_m(t_1, \dots, t_m)$ dominates t_i, that is, $s[j] \geq t_i[j]$ for $1 \leq j \leq m-1$. To show this we count the number of tuples that can fail to be dominated by s. If a tuple t_p is not dominated by s, for some $1 \leq p \leq m$, it means that there is a position $1 \leq j \leq m-1$ such that $t_p[j] > s[j]$. But since Second_m returns the second largest value, for every $1 \leq j \leq m-1$, there is at most one tuple that is not dominated by s. Since there are $m \geq 3$ tuples, there must be an i such that t_i is dominated by s. Moreover, $\text{Max}_m(t_1, \dots, t_m)$ clearly dominates all $t_1, \dots, t_m$. By Proposition 3.9, ϕ is antitone and therefore $\{\langle m-1, \text{Max}_m\rangle, \langle 1, \text{Second}_m\rangle\}$ is a fractional polymorphism of ϕ, by Definition 1.10.   □

Proposition 3.11 *For all $d \geq 2$ and $m \geq 2$, $\{\langle m-1, \text{Max}_m\rangle, \langle 1, \text{Second}_m\rangle\} \notin$* $\text{fPol}(\mathbf{F}_{d,m}^{\max})$.

Proof Let ϕ be the m-ary finite-valued max-closed cost function, over any domain containing $\{0, 1\}$, defined by $\phi(\langle 0, \dots, 0\rangle) = 1$ and $\phi(\langle \cdot, \dots, \cdot\rangle) = 0$ otherwise. To show that $\{\langle m-1, \text{Max}_m\rangle, \langle 1, \text{Second}_m\rangle\}$ is *not* a fractional polymorphism of ϕ, consider the following m-tuples: $\langle 0, \dots, 0, 1\rangle$, $\langle 0, \dots, 0, 1, 0\rangle, \dots, \langle 1, 0, \dots, 0\rangle$. Each of them is assigned cost 0 by ϕ. But applying the functions Max_m $((m-1)$ times) and Second_m coordinatewise results in $m-1$ tuples $\langle 1, \dots, 1\rangle$, which are assigned cost 0 by ϕ, and one tuple $\langle 0, \dots, 0\rangle$, which is assigned cost 1 by ϕ (see Fig. 3.5).   □

Theorem 3.8 *For all $d \geq 2$, $\langle \mathbf{F}_{d,1}^{\max}\rangle \subsetneq \langle \mathbf{F}_{d,2}^{\max}\rangle \subsetneq \langle \mathbf{F}_{d,3}^{\max}\rangle \subsetneq \langle \mathbf{F}_{d,4}^{\max}\rangle \subsetneq \cdots$.*

Proof By Propositions 3.10 and 3.11 and Theorem 1.3.   □

Figure 3.6 summarises the results from this section.

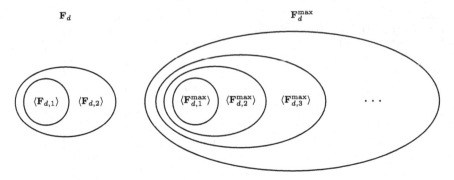

Fig. 3.6 Summary of results from Sect. 3.4, for all $d \geq 2$

3.5 General-Valued Cost Functions

In this section, we show that general-valued cost functions of a fixed arity can express cost functions of arbitrary arities. Comparing this result with the results of the previous section provides a striking example of the way in which allowing infinite cost values in a valued constraint language Γ can drastically affect the expressibility of cost functions over Γ, including finite-valued cost functions.

The class of general-valued max-closed cost functions is known to be tractable [67]. Once again it is straightforward to establish a separation between unary and binary general-valued cost functions.

Proposition 3.12 $\langle \mathbf{G}_{d,1} \rangle \subsetneq \langle \mathbf{G}_{d,2} \rangle$ *and* $\langle \mathbf{G}_{d,1}^{\max} \rangle \subsetneq \langle \mathbf{G}_{d,2}^{\max} \rangle$.

Proof Identical to the proof of Proposition 3.8.   □

As with crisp cost functions, in the special case of a Boolean domain, we can show a separation between binary and ternary general-valued cost functions.

Proposition 3.13 $\langle \mathbf{G}_{2,2} \rangle \subsetneq \langle \mathbf{G}_{2,3} \rangle$ *and* $\langle \mathbf{G}_{2,2}^{\max} \rangle \subsetneq \langle \mathbf{G}_{2,3}^{\max} \rangle$.

Proof By Proposition 3.4, MAJORITY $\in$ Pol($\mathbf{G}_{2,2}$). Moreover, MAJORITY $\in$ Pol($\mathbf{G}_{2,2}^{\max}$). By Proposition 3.5, MAJORITY $\notin$ Pol($\mathbf{G}_{2,3}$) and MAJORITY $\notin$ Pol($\mathbf{G}_{2,3}^{\max}$). The result then follows by Theorem 1.3.   □

Next we show a collapse result for general-valued cost functions.

Theorem 3.9 *For all* $d \geq 3$, $\langle \mathbf{G}_{d,1} \rangle \subsetneq \langle \mathbf{G}_{d,2} \rangle = \mathbf{G}_d$. *Moreover,* $\langle \mathbf{G}_{2,1} \rangle \subsetneq \langle \mathbf{G}_{2,2} \rangle \subsetneq \langle \mathbf{G}_{2,3} \rangle = \mathbf{G}_2$.

Proof Let $\phi \in \mathbf{G}_{d,m}$ for some $d \geq 3$ and $m > 2$. It is easy to check that the same construction as in the proof of Theorem 3.7 can be used to express ϕ, with $K = \infty$.

Now let $\phi \in \mathbf{G}_{2,m}$ for some $m > 2$. It is easy to check that a similar construction to that used in the proof of Theorem 3.7 can be used to express ϕ, where the instance $\mathcal{P}$ is replaced by the ternary Boolean relation that expresses the truth table of a 2-input or-gate. $\qquad \square$

Note that the proof shows a slightly stronger result: $\mathbf{G}_2 = \langle \mathbf{R}_{2,3} \cup \mathbf{F}_{2,1} \rangle$, and for all $d \geq 3$, $\mathbf{G}_d = \langle \mathbf{R}_{d,2} \cup \mathbf{F}_{d,1} \rangle$. In other words, all general-valued cost functions can be expressed using *unary* finite-valued cost functions together with ternary relations (in the case $d = 2$) or binary relations (in the case $d \geq 3$).

By investigating polymorphisms and fractional polymorphisms, we will now show a collapse result for general-valued max-closed cost functions.

First we show that general-valued max-closed cost functions of a fixed arity have the same polymorphisms as max-closed cost functions of arbitrary arities.

Proposition 3.14 *For all $d \geq 3$,*

$$\mathsf{Pol}\big(\mathbf{G}_{d,2}^{\mathrm{max}}\big) = \mathsf{Pol}\big(\mathbf{G}_d^{\mathrm{max}}\big).$$

Moreover, $\mathsf{Pol}(\mathbf{G}_{2,3}^{\mathrm{max}}) = \mathsf{Pol}(\mathbf{G}_2^{\mathrm{max}})$.

Proof Assume, for contradiction, that there is an $f \in \mathsf{Pol}(\mathbf{G}_{d,2}^{\mathrm{max}})$, such that $f \notin \mathsf{Pol}(\mathbf{G}_d^{\mathrm{max}})$. By Definition 3.3, $\{\mathsf{Feas}(\phi) \mid \phi \in \mathbf{G}_d^{\mathrm{max}}\} = \mathbf{R}_d^{\mathrm{max}}$. Therefore, such an f would contradict Theorem 3.6 since $\mathsf{Pol}(\mathbf{R}_{d,2}^{\mathrm{max}}) = \mathsf{Pol}(\mathbf{R}_d^{\mathrm{max}})$.

Similarly, assume that there is an $f \in \mathsf{Pol}(\mathbf{G}_{2,3}^{\mathrm{max}})$ such that $f \notin \mathsf{Pol}(\mathbf{G}_2^{\mathrm{max}})$. This would contradict Theorem 3.4 since $\mathsf{Pol}(\mathbf{R}_{2,3}^{\mathrm{max}}) = \mathsf{Pol}(\mathbf{R}_2^{\mathrm{max}})$. $\qquad \square$

We now prove that general-valued max-closed cost functions of a fixed arity have the same fractional polymorphisms as general-valued max-closed cost functions of arbitrary arities. First we characterise the polymorphisms of general max-closed cost functions.

Proposition 3.15 *For all $d \geq 2$,*

$$\mathsf{Pol}\big(\mathbf{G}_d^{\mathrm{max}}\big) = \big\{\mathrm{Max}_I \mid \emptyset \neq I \subseteq \{1, \ldots, k\}, k = 1, 2, \ldots\big\}.$$

Proof It follows from Definition 3.3 that $\{\mathsf{Feas}(\phi) \mid \phi \in \mathbf{G}_d^{\mathrm{max}}\} = \mathbf{R}_d^{\mathrm{max}}$. Therefore, $\mathsf{Pol}(\mathbf{G}_d^{\mathrm{max}}) = \mathsf{Pol}(\mathbf{R}_d^{\mathrm{max}})$ and the result follows from Proposition 3.2. $\qquad \square$

Next we characterise the fractional polymorphisms of general-valued max-closed cost functions.

Definition 3.6 Let $\mathcal{F} = \{(r_1, \mathrm{MAX}_{S_1}), \ldots, (r_n, \mathrm{MAX}_{S_n})\}$ be a k-ary fractional operation and $S \subseteq \{1, \ldots, k\}$.

We define

$$\mathrm{supp}_{\mathcal{F}} S \overset{\mathrm{def}}{=} \{i \mid S_i \cap S \neq \emptyset\},$$

and

$$\text{wt}_{\mathcal{F}}(S) \overset{\text{def}}{=} \sum_{i \in \text{supp}_{\mathcal{F}}(S)} r_i.$$

Theorem 3.10 *Let* $\mathcal{F} = \{(r_1, \text{MAX}_{S_1}), \ldots, (r_n, \text{MAX}_{S_n})\}$ *be a* k-*ary fractional operation. The following are equivalent*:

1. $\mathcal{F} \in \text{fPol}(\mathbf{G}_d^{\max})$.
2. $\mathcal{F} \in \text{fPol}(\mathbf{G}_{d,1}^{\max})$.
3. *For every subset* $S \subseteq \{1, \ldots, k\}$, $\text{wt}_{\mathcal{F}}(S) \geq |S|$.

Proof We first show that $\neg(3) \Rightarrow \neg(2) \Rightarrow \neg(1)$.

First suppose that there exists an $S \subseteq \{1, \ldots, k\}$ such that $\text{wt}_{\mathcal{F}}(S) < |S|$. Let $\{a, b\} \subseteq D$ be the two largest elements of D and $a < b$. Consider the unary cost function ϕ where

$$\phi(x) \overset{\text{def}}{=} \begin{cases} 0 & \text{if } x = b, \\ 1 & \text{if } x = a, \\ \infty & \text{otherwise.} \end{cases}$$

Certainly $\phi \in \mathbf{G}_{d,1}^{\max}$.

Now let

$$x_i \overset{\text{def}}{=} \begin{cases} b & \text{if } i \in S, \\ a & \text{if } i \notin S. \end{cases}$$

We have that

$$\sum_{i=1}^{k} \phi(x_i) = k - |S|, \quad \text{and}$$

$$\sum_{j=1}^{n} r_j \phi\big(\text{MAX}_{S_j}(x_1, \ldots, x_k)\big) = \sum_{j \notin \text{supp}_{\mathcal{F}} S} r_j \phi(a) + \sum_{j \in \text{supp}_{\mathcal{F}} S} r_j \phi(b)$$

$$= k - \text{wt}_{\mathcal{F}}(S)$$

$$> k - |S|, \quad \text{by assumption.}$$

So $\mathcal{F}$ is not a fractional polymorphism of $\mathbf{G}_{d,1}^{\max}$, and hence not a fractional polymorphism of $\mathbf{G}_d^{\max}$.

To complete the proof we will show that $(3) \Rightarrow (1)$.

Suppose that, for every subset $S \subseteq \{1, \ldots, k\}$, $\text{wt}_{\mathcal{F}}(S) \geq |S|$.

We will first show the existence of a set of nonnegative values p_{ji} for $j = 1, \ldots, n$ and $i = 1, \ldots, k$, where

$$\sum_{i=1}^{k} p_{ji} = r_j,$$

$$\sum_{j=1}^{n} p_{ji} = 1 \quad \text{and}$$

$$p_{ji} = 0 \quad \text{if } i \notin S_j.$$

Consider the network in Fig. 3.7. The capacity from the source to any node x_i is 1. The capacity from node y_j to the sink is r_j. There is an arc from node x_i to node y_j precisely when $i \in S_j$, and the capacity of these arcs is infinite. We show that the flow from x_i to y_j in a maximum flow is the value of p_{ji}.

We will use the (s, t)-MIN-CUT MAX-FLOW Theorem to generate the p_{ji}.

Suppose that we have a minimum cut of this network. Let A be those arcs in this cut from the source to any node x_i. Let $S = \{1, \dots, k\} - \{i \mid x_i \in A\}$. Since we have a cut we must (at least) cut every arc from the nodes $\{y_j \mid j \in \text{supp}_{\mathcal{F}}(S)\}$ to the sink. By assumption, $\text{wt}_{\mathcal{F}}(S) \geq |S|$ and so this cut has total cost at least k. Certainly there is a cut of cost exactly k (cut all arcs from the source), and so the max-flow through this network is precisely k. Such a flow can only be achieved if each arc from the source and each arc to the sink is filled to its capacity. The flow along the arc from x_i to y_j then gives the required value for p_{ji}.

We will use these values p_{ji} to show that $\mathcal{F}$ is indeed a fractional polymorphism of $\mathbf{G}_d^{\max}$.

Let $t_1, \dots, t_k$ be m-ary tuples and $\phi \in \mathbf{G}_{d,m}^{\max}$ be an m-ary cost function. We have to show the following:

$$\sum_{i=1}^{k} \phi(t_i) \geq \sum_{j=1}^{n} r_j \phi\big(\text{MAX}_{S_j}(t_1, \dots, t_k)\big). \tag{3.1}$$

If any $\phi(t_i)$ is infinite, then this inequality clearly holds.

By Proposition 3.15, all MAX_{S_j}, $1 \leq j \leq n$, are polymorphisms of $\mathbf{G}_d^{\max}$. Therefore, if all $\phi(t_i)$ are finite, then all $\phi(\text{MAX}_{S_j}(t_1, \dots, t_k))$ are finite as well.

By definition of p_{ji}, and using the fact that $p_{ji} = 0$ whenever $i \notin S_j$, we have that

$$\sum_{j=1}^{n} r_j \phi\big(\text{MAX}_{S_j}(t_1, \dots, t_k)\big) = \sum_{j=1}^{n} \sum_{i \in S_j} p_{ji} \phi\big(\text{MAX}_{S_j}(t_1, \dots, t_k)\big).$$

Now, since ϕ is antitone, we have

$$\sum_{j=1}^{n} \sum_{i \in S_j} p_{ji} \phi\big(\text{MAX}_{S_j}(t_1, \dots, t_k)\big) \leq \sum_{j=1}^{n} \sum_{i \in S_j} p_{ji} \phi(t_i).$$

Since $p_{ji} = 0$ whenever $i \notin S_j$, we have that

$$\sum_{j=1}^{n} \sum_{i \in S_j} p_{ji} \phi(t_i) = \sum_{j=1}^{n} \sum_{i=1}^{k} p_{ji} \phi(t_i).$$

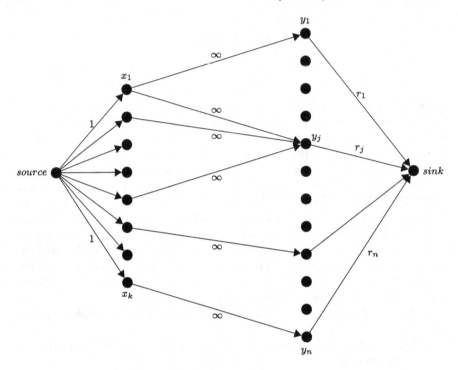

Fig. 3.7 Network for p_{ij}'s in the proof of Theorem 3.10

Finally, since $\sum_{j=1}^{n} p_{ji} = 1$, we have established Inequality (3.1).   □

Theorem 3.11 *For all $d \geq 3$, it holds that* $\mathsf{fPol}(\mathbf{G}_{d,2}^{max}) = \mathsf{fPol}(\mathbf{G}_{d}^{max})$. *Moreover,* $\mathsf{fPol}(\mathbf{G}_{2,3}^{max}) = \mathsf{fPol}(\mathbf{G}_{2}^{max})$.

Proof By Proposition 3.14, $\mathbf{G}_{d,2}^{max}$ and $\mathbf{G}_{d}^{max}$ have the same polymorphisms. Also, $\mathbf{G}_{2,3}^{max}$ and $\mathbf{G}_{2}^{max}$ have the same polymorphisms. By Proposition 3.15, these polymorphisms are of the form "max-on-a-subset". Clearly, each component function of a fractional polymorphism has to be a polymorphism, by Observation 1.2. Therefore, the result follows from Theorem 3.10.   □

Theorem 3.12 *For all $d \geq 3$,* $\langle \mathbf{G}_{d,1}^{max} \rangle \subsetneq \langle \mathbf{G}_{d,2}^{max} \rangle = \mathbf{G}_{d}^{max}$. *Moreover,* $\langle \mathbf{G}_{2,1}^{max} \rangle \subsetneq \langle \mathbf{G}_{2,2}^{max} \rangle \subsetneq \langle \mathbf{G}_{2,3}^{max} \rangle = \mathbf{G}_{2}^{max}$.

Proof The separation results were obtained in Propositions 3.12 and 3.13, by showing that the valued constraint languages involved have different polymorphisms.

For all $d' \geq 2$, $m \geq 1$ and $c \in \mathbb{Q}_{\geq 0}$, $\mathbf{G}_{d',m}^{max}$ is closed under scaling by c. Therefore, using Theorem 1.3, the collapses follow from Proposition 3.14 and Theorem 3.11. □

We now present a non-algebraic proof of the collapse results from Theorem 3.12. In fact, we prove a slightly stronger result: for all $d \geq 3$, $\mathbf{G}_{d}^{max} = \langle \mathbf{R}_{d,2}^{max} \cup \mathbf{F}_{d,1}^{max} \rangle$, and

$\mathbf{G}_2^{\max} = \langle \mathbf{R}_{2,3}^{\max} \cup \mathbf{F}_{2,1}^{\max} \rangle$. However, the following proof requires that cost functions take integer values.

Proof (Alternative proof of Theorem 3.12) Let ϕ be an m-ary general-valued max-closed cost function, and write $x_1, \ldots, x_m$ for the variables. Let $y_1, \ldots, y_K$ be variables (with d values, say $1, \ldots, d$), where $K = \max\{\phi(\mathbf{x}) | \phi(\mathbf{x}) < \infty\}$ is the largest finite cost in the range of ϕ. Intuitively, a cost of k for a tuple will be encoded by $y_1, \ldots, y_k$ assigned the value 1 (and the other variables assigned the value 0).

We first encode infinite costs. Let $\phi_R = \text{Feas}(\phi)$ be the relation containing tuples u such that $\phi(u) < \infty$. It turns out that this relation is max-closed. Indeed, for all $u, v \in \phi_R$ we have $\phi(u), \phi(v) < \infty$ by definition of ϕ_R. Since ϕ is max-closed, we have $2\phi(\text{Max}(u, v)) \leq \phi(u) + \phi(v) < \infty$; so $\phi(\text{Max}(u, v)) < \infty$ and thus $\text{Max}(u, v) \in \phi_R$. So ϕ_R is max-closed, that is, $\phi_R \in \mathbf{R}_d^{\max}$.

We now encode finite costs. For an m-tuple t with $\phi(t) < \infty$, write k_t for $\phi(t)$. We let ψ_t be the anti-Horn formula $\bigwedge_{j=1}^{k_t}((\bigvee_{i=1}^{m} x_i > t[i]) \vee y_j \leq 1)$. Observe that this formula reads $\mathbf{x} \leq t \rightarrow y_1 \leq 1 \wedge \cdots \wedge y_{k_t} \leq 1$.

Finally, we define the anti-Horn formula ψ to be $\psi_R \wedge \bigwedge_{t \in D^m} \psi_t$, where ψ_R is an anti-Horn formula equivalent to ϕ_R. By Proposition 3.6, this formula can be expressed over $\mathbf{R}_{d,2}^{\max}$.

The formula ψ encodes the cost of every tuple as a number of y_j's assigned the value 1. We thus add, to every variable y_j, $j = 1, \ldots, K$, the cost function μ defined by $\mu(1) = 1$ and $\mu(2) = \cdots = \mu(d) = 0$. Clearly, this function is max-closed and therefore in $\mathbf{F}_{d,1}^{\max}$.

We now show that the gadget $\langle \psi, \langle x_1, \ldots, x_m \rangle \rangle$ expresses ϕ. Let t be an m-ary tuple. Assume first that $k_t = \phi(t)$ is finite. Then ψ contains the subformula

$$\bigwedge_{j=1}^{k_t} \left(\left(\bigvee_{i=1}^{m} x_i > t[i] \right) \vee y_j \leq 1 \right).$$

Since obviously $t[i] > t[i]$ holds for no i, every assignment which satisfies ψ sets variables $y_1, \ldots, y_{k_t}$ to 1 and thus has a cost of at least k_t. Now let s_t be the assignment which is equal to t over $x_1, \ldots, x_m$ and which assigns 1 to $y_1, \ldots, y_{k_t}$ and 2 to $y_{k+1}, \ldots, y_K$. We show that s_t satisfies ψ, which gives an assignment of cost at most k_t.

First let $\psi_{t'} \in \psi$, and recall that $\psi_{t'}$ reads $\mathbf{x} \leq t' \rightarrow y_1 \leq 1 \wedge \cdots \wedge y_{\phi(t')} \leq 1$. If $t \leq t'$, then since ϕ is max-closed and both costs are finite (by definition of $\psi_{t'}$), we have $\phi(t) \geq \phi(t')$ by Proposition 3.9. It follows that $\{y_1, \ldots, y_{\phi(t')}\} \subseteq \{y_1, \ldots, y_{k_t}\}$, so s_t assigns 1 to $y_1, \ldots, y_{\phi(t')}$ and thus satisfies $\psi_{t'}$. Otherwise, if $t \not\leq t'$, then $t[i] > t'[i]$ for some i and thus s_t satisfies $\psi_{t'}$ (it does not satisfy its premises). Finally, s_t satisfies $\psi_{t'}$ for all t'.

Now s_t satisfies ψ_R by definition of ϕ_R, since s_t equals t over $x_1, \ldots, x_m$ and $\phi(t) < \infty$ by our assumption. We finally have that for all t with finite cost under ϕ, s_t satisfies ψ, and thus the projection of ψ assigns a cost of at most k_t to t. Since the cost is at least k_t as shown above, we are done.

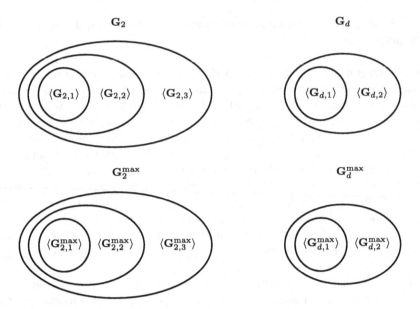

Fig. 3.8 Summary of results from Sect. 3.5, for all $d \geq 3$

Now if t has infinite cost under ϕ, then by definition of ϕ_R we have that t does not satisfy ψ and thus has infinite cost. Therefore, for all $d \geq 3$, $\mathbf{G}_d^{\max} = \langle \mathbf{R}_{d,2}^{\max} \cup \mathbf{F}_{d,1}^{\max} \rangle$.

In the case of a Boolean domain, ψ is a relation over the Boolean domain, and therefore ψ can be expressed, by Proposition 3.6, over $\mathbf{R}_{2,3}^{\max}$. Therefore, $\mathbf{G}_2^{\max} = \langle \mathbf{R}_{2,3}^{\max} \cup \mathbf{F}_{2,1}^{\max} \rangle$. □

Figure 3.8 summarises the results from this section.

Example 3.1 Consider the ternary finite-valued max-closed cost function ϕ over $D = \{0, 1, 2\}$ defined by

$$\phi(t) \stackrel{\text{def}}{=} \begin{cases} 1 & \text{if } t = \langle 0, 0, 0 \rangle, \\ 0 & \text{otherwise.} \end{cases}$$

By Proposition 3.11, $\phi \notin \langle \mathbf{F}_{3,2}^{\max} \rangle$. In other words, ϕ is not expressible using only finite-valued max-closed cost functions of arity at most 2. However, by Theorem 3.12, $\phi \in \langle \mathbf{G}_{3,2}^{\max} \rangle$. We now show how ϕ can be expressed using general-valued max-closed cost functions of arity at most 2.

Let ϕ_0 be the binary finite-valued max-closed cost function defined as follows:

$$\phi_0(t) \stackrel{\text{def}}{=} \begin{cases} 1 & \text{if } t = \langle 0, 0 \rangle, \\ 0 & \text{otherwise.} \end{cases}$$

Fig. 3.9 $\mathcal{P}_1$, an instance of VCSP($\mathbf{G}_{3,2}^{\max}$) expressing ϕ (Example 3.1)

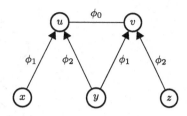

Next, define two binary crisp[2] max-closed cost functions

$$\phi_1(t) \overset{\text{def}}{=} \begin{cases} \infty & \text{if } t = \langle 0, 1 \rangle, \\ 0 & \text{otherwise,} \end{cases}$$

and

$$\phi_2(t) \overset{\text{def}}{=} \begin{cases} \infty & \text{if } t = \langle 0, 2 \rangle, \\ 0 & \text{otherwise.} \end{cases}$$

Let $\mathcal{P}_1 = \langle V, D, C \rangle$, where $V = \{x, y, z, u, v\}$ and

$$C = \{ \langle \langle x, u \rangle, \phi_1 \rangle, \langle \langle y, u \rangle, \phi_2 \rangle, \langle \langle y, v \rangle, \phi_1 \rangle, \langle \langle z, v \rangle, \phi_2 \rangle, \langle \langle u, v \rangle, \phi_0 \rangle \}.$$

We claim that $\langle \mathcal{P}_1, \langle x, y, z \rangle \rangle$ is a gadget for expressing ϕ over $\mathbf{G}_{3,2}^{\max}$. (See Fig. 3.9.) If any of x, y, z is nonzero, then at least one of the variables u, v can be assigned a nonzero value and the cost of such an assignment is 0. Conversely, if x, y, and z are all assigned 0, then the minimum-cost assignment must also assign 0 to both u and v, and hence has cost 1.

We now show another gadget for expressing ϕ, using only crisp max-closed cost functions of arity at most 2 and finite-valued max-closed cost functions of arity at most 1.

Fig. 3.10 $\mathcal{P}_2$, an instance of VCSP($\mathbf{R}_{3,2}^{\max} \cup \mathbf{F}_{3,1}^{\max}$) expressing ϕ (Example 3.1)

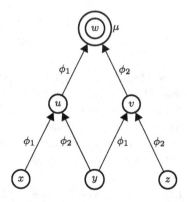

[2]Note that a "finite variant" of ϕ_1, defined by $\phi_1(\langle 0, 1 \rangle) = K$ for some finite $K < \infty$ and $\phi_1(\langle \cdot, \cdot \rangle) = 0$ otherwise, is not max-closed. The infinite cost is necessary.

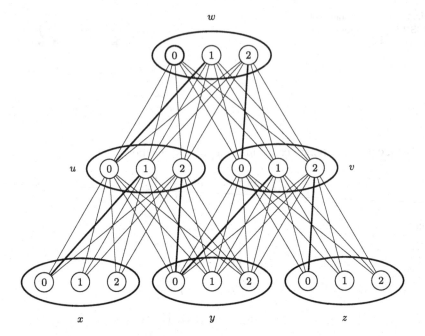

Fig. 3.11 Microstructure of the instance $\mathcal{P}_2$ (Example 3.1)

Let μ be the unary finite-valued max-closed cost function defined by

$$\mu(x) \stackrel{\text{def}}{=} \begin{cases} 1 & \text{if } x = 0, \\ 0 & \text{otherwise.} \end{cases}$$

Let $\mathcal{P}_2 = \langle V', D, \mathcal{C}' \rangle$, where $V' = \{x, y, z, u, v, w\}$ and

$$\mathcal{C}' = \{ \langle \langle x, u \rangle, \phi_1 \rangle, \langle \langle y, u \rangle, \phi_2 \rangle, \langle \langle y, v \rangle, \phi_1 \rangle,$$
$$\langle \langle z, v \rangle, \phi_2 \rangle, \langle \langle u, w \rangle, \phi_1 \rangle, \langle \langle v, w \rangle, \phi_2 \rangle, \langle w, \mu \rangle \}.$$

See Fig. 3.10. Applying an argument similar to the one above, it can be shown that $\langle \mathcal{P}_2, \langle x, y, z \rangle \rangle$ is a gadget for expressing ϕ. This can be verified by examining the microstructure of $\mathcal{P}_2$ (see Fig. 3.11): circles represent particular assignments to particular variables, as indicated, and edges are weighted by the cost of the corresponding pairs of assignments. Thin edges indicate zero weight, bold edges indicate infinite weight, and assigning 0 to variable w gives cost 1.

Example 3.2 Recall Example 1.30, which showed how to express $\phi = (\#0)^2$ using cost functions ϕ_0, ϕ_1, ϕ_2, and μ. All three binary crisp cost functions ϕ_0, ϕ_1, and ϕ_2 are max-closed. Moreover, the unary finite-valued cost function μ is max-closed as well.

3.6 Characterisation of Mul($\mathbf{F}_d^{\max}$) and fPol($\mathbf{F}_d^{\max}$)

In Sect. 3.4, we have seen that fPol($\mathbf{F}_{d,m+1}^{\max}$) $\nsubseteq$ fPol($\mathbf{F}_{d,m}^{\max}$) implies an infinite hierarchy of cost functions with ever-increasing expressive power. In Sect. 3.5, we have characterised the fractional clone of general-valued max-closed cost functions. In this section, we characterise the multimorphisms and fractional polymorphisms of finite-valued max-closed cost functions.

Notation 3.2 Recall from Notation 3.1 that we say that an m-tuple u *dominates* an m-tuple v, denoted by $u \geq v$, if $u[i] \geq v[i]$ for all $1 \leq i \leq m$. We say that u *strictly dominates* v if $u[i] > v[i]$ for all $1 \leq i \leq m$. If $s \geq t$ ($s > t$), we also write $t \leq s$ ($t < s$).

Notation 3.3 For a graph $G = (V, E)$ and a set of vertices $V' \subseteq V$, we define the set of neighbours of V' as $N(V') = \{v \in V | (\exists v' \in V) \ [(v, v') \in E]]\}$. We say a graph $G = (V, E)$ is *bipartite* if $V = V_0 \dot{\cup} V_1$ and $E \subseteq V_0 \times V_1$. For a bipartite graph $G = (V_0, V_1, E)$, a *matching* is a set of edges $E' \subseteq E$ such that no two edges from E' share a vertex, and the size of such a matching is $|E'|$. A *perfect matching* is a matching of size $|V_0|$.

Theorem 3.13 (Hall (1935)) *A bipartite graph $G = (V_0, V_1, E)$ has a perfect matching if, and only if, $(\forall V' \subseteq V_0) \ [|N(V')| \geq |V'|]$.*

We denote by D a fixed finite totally-ordered domain set with $|D| = d$.

We now characterise the multimorphisms of $\mathbf{F}_d^{\max}$, the valued constraint language containing all finite-valued max-closed cost functions.

Theorem 3.14 *Let $\mathcal{F} = \langle f_1, \ldots, f_k \rangle$, where each $f_i : D^k \to D$, and let $S_i = \{j \mid f_i(u_1, \ldots, u_k) \geq u_j\}$, $1 \leq i \leq k$. Then the following are equivalent:*

1. *$\mathcal{F} \in$ Mul($\mathbf{F}_d^{\max}$).*
2. *For every fixed collection of k tuples $u_1, \ldots, u_k \in D^m$,*

$$\left(\forall I \subseteq \{1, \ldots, k\} \right) \left[\left| \bigcup_{i \in I} S_i \right| \geq |I| \right].$$

3. *For every fixed collection of k tuples $u_1, \ldots, u_k \in D^m$, there exists a bijective mapping $\varphi : \{1, \ldots, k\} \to \{1, \ldots, k\}$ such that*

$$(\forall 1 \leq i \leq k) \left[f_i(u_1, \ldots, u_k) \geq u_{\varphi(i)} \right].$$

Proof First we show (1) $\Rightarrow$ (2).

Let $I \subseteq \{1, \ldots, k\}$ and assume for contradiction that $|\bigcup_{i \in I} S_i| < |I|$. Define a cost function ϕ on D^m as

$$\phi(x) \stackrel{\text{def}}{=} \begin{cases} 1 & \text{if } x \leq f_i(u_1, \ldots, u_k) \text{ for some } i \in I, \\ 0 & \text{otherwise.} \end{cases}$$

We show that ϕ is max-closed. Let u and v be two m-tuples. If $\phi(u) = \phi(v) = 1$, then the multimorphism inequality, as given in Definition 1.8, is satisfied easily as ϕ takes only costs 0 and 1. If either $\phi(u) = 0$ or $\phi(v) = 0$ (or both), then $\phi(\text{Max}(u, v)) = 0$ from the definition of ϕ, and the multimorphism inequality is satisfied again. Hence ϕ is max-closed.

The following holds from the definition of ϕ:

$$\sum_{i=1}^{k} \phi(u_i) = \left| \bigcup_{i \in I} S_i \right| < |I| \leq \sum_{i}^{k} \phi(f_i(u_1, \ldots, u_k)).$$

Therefore, $\mathcal{F}$ is not a multimorphism of ϕ, which is a contradiction as ϕ is a finite-valued max-closed cost function.

Now we show (2) $\Rightarrow$ (3).

Consider a bipartite graph $G = (V_0, V_1, E)$, where

$$V_0 = \{0\} \times \{f_1(u_1, \ldots, u_k), \ldots, f_k(u_1, \ldots, u_k)\},$$
$$V_1 = \{1\} \times \{u_1, \ldots, u_k\},$$

and

$$\left(\langle 0, f_i(u_1, \ldots, u_k) \rangle, \langle 1, u_j \rangle \right) \in E \quad \Leftrightarrow \quad (j \in S_i).$$

(Since $\{u_1, \ldots, u_k\}$ and $\{f_1(u_1, \ldots, u_k), \ldots, f_k(u_1, \ldots, u_k)\}$ are not necessarily disjoint, we distinguish between the same element from $\{u_1, \ldots, u_k\}$ and $\{f_1(u_1, \ldots, u_k), \ldots, f_k(u_1, \ldots, u_k)\}$.) We have that for any $V' \subseteq V_0$, $N(V') = \bigcup_{\{i \mid \langle 0, f_i(u_1, \ldots, u_k) \rangle \in V'\}} S_i$. Therefore, (2) is equivalent to Hall's Condition, and by Hall's Theorem 3.13, G has a perfect matching. This matching clearly defines the wanted bijective mapping φ.

Finally, we show (3) $\Rightarrow$ (1).

This is clear as finite-valued max-closed cost functions are finitely antitone by Proposition 3.9, and therefore the multimorphism inequality, as given in Definition 1.8, is satisfied. $\qquad \square$

In other words, a function $\mathcal{F}: D^k \to D^k$ is a multimorphism of $\mathbf{F}_d^{\max}$ if, and only if, for every collection of source k tuples, there is a bijective mapping that maps the k source tuples to the k target tuples in a dominance-preserving way. Note that the same proof characterises the multimorphisms of $\mathbf{F}_{d,m}^{\max}$ for every $m \geq 1$.

Corollary 3.1 $\mathcal{F} \in \text{Mul}(\mathbf{F}_{d,m}^{\max})$ *if, and only if, for every collection of k tuples* $u_1, \ldots, u_k \in D^m$ *there is a bijective mapping* $\varphi: \{1, \ldots, k\} \to \{1, \ldots, k\}$ *such that* $(\forall 1 \leq i \leq k) [f_i(u_1, \ldots, u_k) \geq u_{\varphi(i)}]$.

Proof Notice that ϕ from the proof of Theorem 3.14 is of arity m. Therefore, $\phi \in \mathbf{F}_{d,m}^{\max}$ and the same proof works. $\qquad \square$

The next theorem shows that for a given $\mathcal{F} \in \mathsf{Mul}(\mathbf{F}_d^{\max})$, there is a uniform bijective mapping from Theorem 3.14(3), that is, a bijective mapping that works for every possible collection of tuples.

Theorem 3.15 *For a given* $\mathcal{F} = \langle f_1, \ldots, f_k \rangle \in \mathsf{Mul}(\mathbf{F}_d^{\max})$, *where each* $f_i : D^k \to D$, $1 \leq i \leq k$, *there is a bijective mapping* $\varphi : \{1, \ldots, k\} \to \{1, \ldots, k\}$ *such that for every collection of* k *tuples* $u_1, \ldots, u_k \in D^m$, $(\forall 1 \leq i \leq k)\, [f_i(u_1, \ldots, u_k) \geq u_{\varphi(i)}]$.

Proof By Theorem 3.14, for every collection of k tuples $u_1, \ldots, u_k \in D^m$, there is a bijective mapping with the required properties. Assume for contradiction that there is no single mapping that preserves dominance for every collection of k m-tuples. Define a finite-valued max-closed cost function of arity md^{mk} as

$$\phi(x) \stackrel{\text{def}}{=} \psi(x_1, \ldots, x_m) + \psi(x_{m+1}, \ldots, x_{2m}) + \cdots$$

$$+ \psi(x_{(d^{mk}-1)m+1}, \ldots, x_{d^{mk}m})$$

for some $\psi \in \mathbf{F}_{d,m}^{\max}$. Let u and v be two tuples of arity md^{mk}. Since ψ is max-closed, if $u \leq v$, then $\phi(u) \geq \phi(v)$. Hence ϕ is a max-closed cost function. Therefore, $\mathcal{F}$ is a multimorphism of ϕ. By Theorem 3.14, there is a bijective dominance-preserving mapping for a collection of k (md^{mk})-tuples that corresponds to every possible collection of k m-tuples.   □

In a similar way, we can characterise the fractional polymorphisms of $\mathbf{F}_d^{\max}$.

Theorem 3.16 *Let* $\mathcal{F} = \{\langle r_1, f_1 \rangle, \ldots, \langle r_n, f_n \rangle\}$, *where each* r_i *is a positive rational number such that* $\sum_{i=1}^n r_i = k$ *and each* $f_i : D^k \to D$. *Let* $S_i = \{j \mid f_i(u_1, \ldots, u_k) \geq u_j\}$, $1 \leq i \leq k$. *Then the following are equivalent:*

1. $\mathcal{F} \in \mathsf{fPol}(\mathbf{F}_d^{\max})$.
2. *For every fixed collection of* k *tuples* $u_1, \ldots, u_k \in D^m$,

$$\left(\forall I \subseteq \{1, \ldots, n\}\right) \left[\left|\bigcup_{i \in I} S_i\right| \geq \sum_{i \in I} r_i\right].$$

Proof First we show (1) $\Rightarrow$ (2).

Let $I \subseteq \{1, \ldots, n\}$ and assume for contradiction that $|\bigcup_{i \in I} S_i| < \sum_{i \in I} r_i$. Define a cost function ϕ on D^m as

$$\phi(x) \stackrel{\text{def}}{=} \begin{cases} 1 & \text{if } x \leq f_i(u_1, \ldots, u_k) \text{ for some } i \in I, \\ 0 & \text{otherwise.} \end{cases}$$

We show that ϕ is max-closed. Let u and v be two m-tuples. If $\phi(u) = \phi(v) = 1$, then the multimorphism inequality, as given in Definition 1.8, is satisfied. If either $\phi(u) = 0$ or $\phi(v) = 0$ (or both), then $\phi(\text{Max}(u, v)) = 0$ from the definition of ϕ, and the multimorphism inequality is satisfied again. Hence ϕ is max-closed.

The following holds from the definition of ϕ:

$$\sum_{i=1}^{k} \phi(u_i) = \bigcup_{i \in I} S_i < \sum_{i \in I} r_i \leq \sum_{i=1}^{k} r_i \phi(f_i(u_1, \ldots, u_k)).$$

Therefore, $\mathcal{F}$ is not a fractional polymorphism of ϕ, which is a contradiction as ϕ is a finite-valued max-closed cost function.

Now we show (2) $\Rightarrow$ (1). Let $u_1, \ldots, u_k \in D^m$ be a fixed collection of k tuples. We want to show that if (2) holds, then

$$\sum_{i=1}^{k} \phi(u_i) \geq \sum_{i=1}^{n} r_i \phi(f_i(u_1, \ldots, u_k)). \tag{3.2}$$

Since all r_i, $1 \leq i \leq n$, are rational numbers, we have $r_i = p_i/q_i$. Let $q = \mathrm{lcm}(q_1, \ldots, q_n)$.[3] Then for every $1 \leq i \leq n$, $r_i = ((p_iq)/q_i)(1/q) = k_i(1/q)$, where k_i is a natural number.

For every $1 \leq i \leq n$, we replace the tuple $f_i(u_1, \ldots, u_k)$, which has weight r_i, with k_i copies of the same tuple where each new tuple has weight $1/q$. Since $\sum_{i=1}^{n} r_i = k$, we have kq new tuples and we denote them $v'_1, \ldots, v'_{kq}$. Clearly,

$$\sum_{i=1}^{n} r_i \phi(f_i(u_1, \ldots, u_k)) = \sum_{i=1}^{kq} (1/q)\phi(v'_i).$$

For every $1 \leq i \leq k$, we replace the tuple u_i, which has (implicit) weight 1, with q copies of the same tuple where each new tuple has weight $1/q$. We denote these new tuples $u'_1, \ldots, u'_{kq}$. Clearly,

$$\sum_{i=1}^{k} \phi(u_i) = \sum_{i=1}^{kq} (1/q)\phi(u'_i).$$

Using an idea similar to the one in the proof of Theorem 3.14, consider a bipartite graph $G = (V_0, V_1, E)$, where $V_0 = \{v'_1, \ldots, v'_{kq}\}$, $V_1 = \{u'_1, \ldots, u'_{kq}\}$, and $(v'_{i'}, u'_{j'}) \in E \Leftrightarrow v'_{i'}$ replaced some $f_i(u_1, \ldots, u_k)$, and $u'_{j'}$ replaced some u_j, $j \in S_i$.

Clearly, (2) implies Hall's Condition on G. By Theorem 3.13, G has a perfect matching. As edges in G preserve dominance on tuples, and every finite-valued max-closed cost function is antitone by Proposition 3.9, Inequality (3.2) is satisfied. Therefore, $\mathcal{F} \in \mathrm{fPol}(\mathbf{F}_d^{\mathrm{max}})$.   □

---

[3]Least common multiple.

3.7 Summary

We have investigated the expressive power of valued constraints in general and max-closed valued constraints in particular.

In the case of relations, we built on previously known results about the expressibility of an arbitrary relation in terms of binary or ternary relations. We were able to prove, in a similar way, that an arbitrary max-closed relation can be expressed using binary or ternary max-closed relations. The results about the collapse of the set of all relations and all max-closed relations contrast sharply with the case of finite-valued max-closed cost functions, where we showed an infinite hierarchy. This shows that the VCSP is not just a minor generalisation of the CSP—finite-valued max-closed cost functions behave very differently from crisp max-closed cost functions with respect to expressive power. We also showed the collapse of general-valued cost functions, by characterising the polymorphisms and fractional polymorphisms of general-valued max-closed cost functions. This shows that allowing infinite costs in max-closed cost functions increases their expressive power substantially, and sometimes allows more finite-valued functions to be expressed.

All of our results about max-closed cost functions obviously have equivalent versions for *min-closed* cost functions, that is, those that have the fractional polymorphism $\{\langle 2, \text{Min} \rangle\}$. In the Boolean crisp case these are precisely the relations that can be expressed by a conjunction of *Horn* clauses.

Peter Jonsson observed that the proof of the infinite hierarchy result, which we proved in Theorem 3.2 (3) for max-closed cost functions, in fact works for a slightly larger class of cost functions admitting a fractional polymorphism $\{\langle k, \text{Max}_t \rangle\}$, where $\text{Max}_t(a_1, \ldots, a_k) = t$ if $\text{Max}(a_1, \ldots, a_k) \leq t$ and $\text{Max}_t(a_1, \ldots, a_k) = \text{Max}(a_1, \ldots, a_k)$ otherwise.

3.7.1 Related Work

We remark on the relationship between our results and some previous work on the VCSP. Larrosa and Dechter have shown [202] that both the so-called *dual* representation [98] and the *hidden variable* representation [95], which transform any CSP instance into a binary CSP instance, can be generalised to the VCSP framework. However, these representations involve an exponential blowup (in the arity of the constraints) of the domain size (that is, the set of possible values for each variable). The notion of expressibility that we are using in this book always preserves the domain size. Our results clarify which cost functions can be expressed using a given valued constraint language over the same domain, by introducing additional (hidden) variables and constraints; the number of these that are required is fixed for any given cost function.

The class of relations that can be made max-closed by permuting domains of all variables is called *renamable*, *permutable*, or *switchable* anti-Horn. Testing whether a given VCSP instance consists of renamable max-closed cost functions can be done in polynomial time for Boolean domains [204], but is NP-complete for non-Boolean domains [138].

3.7.2 Open Problems

What other classes of cost functions are expressible by cost functions of fixed arities? What other hierarchies of strictly increasing expressive power can be identified?

Chapter 4
Expressibility of Submodular Languages

An algorithm must be seen to be believed.
Donald Knuth

4.1 Introduction

In this chapter, we study the expressive power of binary submodular functions. Our results present known and new classes of submodular functions that are expressible by binary submodular functions.

Recall that SFM_b is the minimisation problem for locally-defined submodular functions. There is a close relationship between the expressive power of binary submodular functions and solving the SFM_b problem via (s, t)-MIN-CUT: showing that a class C of submodular functions is expressible by binary submodular functions is equivalent to showing that the SFM_b problem for functions from C can be solved via (s, t)-MIN-CUT.

As SFM_b is equivalent to VCSP instances with bounded-arity submodular constraints, our results have important consequences for submodular VCSP instances. We will present our results primarily in the language of pseudo-Boolean optimisation. However, in Chap. 5, we will mention the consequences of our results for VCSPs and certain optimisation problems arising in computer vision. This chapter is based on [280].

4.2 Results

Recall from Sect. 1.6 that an instance of the SFM problem can be minimised in polynomial time. The time complexity of the fastest known general algorithm for SFM, and therefore for VCSP instances with submodular constraints, is $O(n^6 + n^5 L)$, where n is the number of variables and L is the look-up time (needed to evaluate the cost of an assignment to all variables) [227].

As discussed in more detail in Sect. 4.3, we will only deal with Boolean instances of SFM_b. Therefore, an instance of SFM_b with n variables will be represented as a polynomial in n Boolean variables, of some fixed bounded degree. The problem of expressing Boolean submodular functions by binary submodular functions is then

S. Živný, *The Complexity of Valued Constraint Satisfaction Problems*,
Cognitive Technologies, DOI 10.1007/978-3-642-33974-5_4,
© Springer-Verlag Berlin Heidelberg 2012

equivalent to expressing Boolean submodular polynomials by binary submodular polynomials.

A general algorithm for SFM can always be used for the more restricted SFM$_b$, but the special features of this more restricted problem sometimes allow more efficient special-purpose algorithms to be used. (Note that we are focusing on *exact* algorithms that find an optimal solution. See [59] for approximation algorithms for the MAX-CSP, which is a special case of the VCSP, and [113] for approximation algorithms for the SFM.) In particular, it has been shown that certain cases can be solved much more efficiently by reducing them to the (s, t)-MIN-CUT problem, that is, the problem of finding a minimum cut in a directed graph which includes a given source vertex and excludes a given target vertex. For example, it has been known since 1965 that the minimisation of *quadratic* submodular polynomials is equivalent to finding a minimum cut in a corresponding directed graph [34, 84, 147]. Hence quadratic submodular polynomials can be minimised in $O(n^3)$ time, where n is the number of variables.

A Boolean polynomial in at most two variables has degree at most 2, so any *sum* of binary Boolean polynomials has degree at most 2; in other words, it is quadratic. It follows that an efficient algorithm, based on reduction to (s, t)-MIN-CUT, can be used to minimise any class of functions that can be written as a sum of binary submodular polynomials. We will say that a polynomial that can be written in this way, perhaps with additional variables to be minimised over, is *expressible* by binary submodular polynomials (cf. Sect. 4.3). The following classes of functions have all been shown to be expressible by binary submodular polynomials in this way[1]:

- polynomials where all terms of degree 2 or more have negative coefficients (also known as *negative-positive* polynomials) [245];
- cubic submodular polynomials [20];
- $\{0, 1\}$-valued submodular functions (also known as 2-monotone functions) [59, 87].

All these classes of functions have been shown to be expressible by binary submodular polynomials and hence to be minimisable in cubic time (in the total number of variables). Moreover, several classes of submodular functions over non-Boolean domains have also been shown to be expressible by binary submodular functions and hence to be minimisable in cubic time [49, 58, 59].

Our results are twofold. First, we provide alternative, and often much simpler, proofs of the expressibility results for the above classes of functions. Second, we present a new class of submodular functions of arbitrary arities expressible by binary submodular polynomials, and hence minimisable in cubic time (in the total number of variables).

This chapter is organised as follows. Section 4.3 provides necessary information on submodular functions. In Sect. 4.4, we show equivalence between the (s, t)-

---

[1] In fact, it is known that *all* Boolean polynomials (of arbitrary degree) are expressible by binary polynomials [34, 247], but the general construction does not preserve submodularity, that is, the resulting binary polynomials are not necessarily submodular.

MIN-CUT problem and the minimisation problem for quadratic submodular polynomials. In Sect. 4.5, we present alternative proofs of known expressibility results for so-called negative-positive, $\{0, 1\}$-valued, and ternary submodular functions. In Sect. 4.6, we present a new class of submodular functions of arbitrary arities expressible by binary submodular functions.

4.3 Preliminaries

Recall that a cost function of arity n is just a mapping from D^n to $\overline{\mathbb{Q}}_{\geq 0}$ for some fixed finite set D. Cost functions can be added and multiplied by arbitrary positive real values; hence for any given set of cost functions, Γ, we define the convex cone generated by Γ, as follows.

Definition 4.1 (Cone) For any set of cost functions Γ, the *cone generated by* Γ, denoted by $\mathsf{Cone}(\Gamma)$, is defined by:

$$\mathsf{Cone}(\Gamma) \stackrel{\text{def}}{=} \{\alpha_1\phi_1 + \cdots + \alpha_r\phi_r \mid r \geq 1; \phi_1, \ldots, \phi_r \in \Gamma; \alpha_1, \ldots, \alpha_r \geq 0\}.$$

Note that Definition 1.4 of expressibility can be stated equivalently as follows:

Definition 4.2 (Expressibility) A cost function ϕ of arity n is said to be *expressible* by a set of cost functions Γ if ϕ can be written as $\phi = \min_{y_1,\ldots,y_j} \phi'(x_1, \ldots, x_n, y_1, \ldots, y_j)$, for some $\phi' \in \mathsf{Cone}(\Gamma)$. The variables $y_1, \ldots, y_j$ are called *extra* (or *hidden*) variables, and ϕ' is called a *gadget* for ϕ over Γ.

Remark 4.1 Recall from Definition 1.5 and Theorem 1.3 that we care about expressibility up to additive and multiplicative constants.

Lemma 4.1 ([67]) *Let D be a finite lattice-ordered set. A cost function $\phi : D^m \to \overline{\mathbb{Q}}_{\geq 0}$ is submodular if, and only if, the following two conditions are satisfied:*

1. ϕ *satisfies that for all m-tuples u, v with $\phi(u), \phi(v) < \infty$,*

$$\phi(\mathsf{Min}(u, v)) + \phi(\mathsf{Max}(u, v)) \leq \phi(u) + \phi(v).$$

2. $\mathsf{Min}, \mathsf{Max} \in \mathsf{Pol}(\{\phi\})$.

The second condition in Lemma 4.1 implies that the set of m-tuples on which ϕ is finite is a sublattice of the set of all m-tuples, where the lattice operations are the operations Min and Max. Theorem 49.2 of [258] proves that any submodular function defined on such a sublattice can be extended to a submodular function defined on the full lattice.[2] Hence, by Lemma 4.1, any submodular function ϕ can be

---

[2]Note that this result is not obvious because simply changing the infinite cost to some big, but finite constant M does not work: for $c_1 < c_2$, $\infty + c_1 \geq \infty + c_2$, but $M + c_1 < M + c_2$. For instance,

expressed as the sum of a finite-valued submodular function ϕ_{fin}, and a submodular relation $\phi_{\text{rel}} = \text{Feas}(\phi)$, that is, $\phi = \phi_{\text{fin}} + \phi_{\text{rel}}$.

It is known that all submodular *relations* are binary decomposable (that is, equal to the sum of their binary projections) [162], and hence are expressible using only binary submodular relations. Therefore, when considering which cost functions are expressible by binary submodular cost functions, we can restrict our attention to *finite-valued* cost functions without any loss of generality.

Remark 4.2 We discuss more the restriction to finite-valued submodular cost functions. Given a finite lattice-ordered set D, let ϕ be a submodular cost function defined on a sublattice D' of D. The goal is to change the definition of f to the whole of D so that the resulting cost function is submodular, and the minimisation problem is not affected by these changes. In other words, we would like to find a finite-valued submodular $\bar{f}$ such that $f = \bar{f}$ on D', $\min f = \min \bar{f}$, and the minimum is achieved on D'.

Schrijver has shown [258] that for a given ϕ as above, there is an $\alpha \in \mathbb{Q}_{\geq 0}$ such that $\bar{f}(u) = f(\bar{u}) + \alpha |\bar{u} \bigtriangleup u|$ is a finite-valued submodular cost function, where $\bar{u}$ is the smallest element above u such that $f(\bar{u}) < \infty$, and $\bar{u} \bigtriangleup u$ is the symmetric difference between the sets corresponding to $\bar{u}$ and u in the 0-1 representation of D. Clearly, the same holds for every $\alpha' \geq \alpha$. For example, in a VCSP instance with submodular valued constraints over n variables, it is sufficient to choose $\alpha' \geq nM$, where M is the maximum finite cost of all cost functions.

We have shown that when dealing with the expressibility problem for submodular cost functions, we can restrict ourselves to only finite-valued cost functions without any loss of generality. Now we show that we can restrict ourselves to only *Boolean* finite-valued cost functions.

Remark 4.3 Any variable over a non-Boolean domain $D = \{0, 1, \ldots, d - 1\}$ of size d, where $d > 2$, can be encoded by $d - 1$ Boolean variables. This process is known as *Booleanisation*. One such encoding is the following: $en(i) = 0^{d-i-1}1^i$. Using this encoding function we can replace each variable with $d - 1$ new Boolean variables and impose a (submodular) relation on these new variables that ensures that they only take values in the range of the encoding function en. Note that $en(\max(a, b)) = \max(en(a), en(b))$ and $en(\min(a, b)) = \min(en(a), en(b))$, so this encoding preserves submodularity. Therefore, in this chapter, we will focus on problems over Boolean domains, that is, where $D = \{0, 1\}$.

Remark 4.4 Because every chain in a lattice-ordered set has to be mapped to a chain, and each chain have to be mapped to a different chain, any submodularity-preserving encoding of a non-Boolean variable over a d-element domain by Boolean variables

---

consider the submodular cost function ϕ defined as follows: $\phi(0, 0) = \phi(1, 0) = \infty$, $\phi(0, 1) = 1$, and $\phi(1, 1) = 2$. Changing $\phi(0, 0) = \phi(1, 0) = M$ for any finite number M would violate the submodularity condition.

needs at least d variables. However, for practical purposes, certain subclasses of non-Boolean submodular functions that can be encoded by Boolean submodular functions with fewer variables have been studied, as well as approximation algorithms for these problems [182, 241].

Any cost function of arity m can be represented as a table of values of size D^m. Moreover, a finite-valued cost function $\phi : D^m \to \mathbb{Q}_{\geq 0}$ on a Boolean domain $D = \{0, 1\}$ can also be represented as a *polynomial* in m (Boolean) variables with coefficients from $\mathbb{Q}$, where the degree of this polynomial is at most m (such functions are sometimes called *pseudo-Boolean functions* [34, 84]). Over a Boolean domain, we have $x^2 = x$, so the degree of any variable in any term can be restricted to 0 or 1, and this polynomial representation is then unique. Hence, in what follows, we will often refer to a finite-valued cost function on a Boolean domain and its corresponding polynomial interchangeably.

For polynomials over Boolean variables there is a standard way to define *derivatives* of each order (see [34, 84]). For example, the second-order derivative of a polynomial p, with respect to the first two indices, denoted by $\delta_{12}(\mathbf{x})$, is defined by $p(1, 1, \mathbf{x}) - p(1, 0, \mathbf{x}) - p(0, 1, \mathbf{x}) + p(0, 0, \mathbf{x})$. Derivatives for other pairs of indices are defined analogously. It has been shown in [224] that a polynomial $p(x_1, \ldots, x_n)$ over Boolean variables $x_1, \ldots, x_n$ represents a submodular cost function if, and only if, its second-order derivatives $\delta_{ij}(\mathbf{x})$ are nonpositive for all $1 \leq i < j \leq n$ and all $\mathbf{x} \in D^{n-2}$. An immediate corollary is that a quadratic polynomial represents a submodular cost function if, and only if, the coefficients of all quadratic terms are nonpositive.

Remark 4.5 Note that a cost function is called *supermodular* if all its second-order derivatives are nonnegative. Clearly, f is submodular if, and only if, $-f$ is supermodular, so it is straightforward to translate results about supermodular functions, such as those given in [59] and [236], into similar results for submodular functions, and we will use this observation several times below. Cost functions which are both submodular and supermodular (in other words, all second-order derivatives are equal to 0) are called *modular*, and polynomials corresponding to modular cost functions are linear [34, 84].

Example 4.1 For any set of indices $I = \{i_1, \ldots, i_m\} \subseteq \{1, \ldots, n\}$ we can define a cost function ϕ_I in n variables as follows:

$$\phi_I(x_1, \ldots, x_n) \overset{\text{def}}{=} \begin{cases} -1 & \text{if } (\forall i \in I)(x_i = 1), \\ 0 & \text{otherwise.} \end{cases}$$

The polynomial representation of ϕ_I is

$$p(x_1, \ldots, x_n) = -x_{i_1} \ldots x_{i_m},$$

which is a polynomial of degree m. By considering the second-order derivatives of p, it follows that ϕ_I is submodular.

However, the function ϕ_I is also expressible by *binary* polynomials, using a single extra variable, y, as follows:

$$\phi_I(x_1, \ldots, x_n) = \min_{y \in \{0,1\}} y\left(m - 1 - \sum_{i \in I} x_i\right).$$

This is a special case of the expressibility result for negative-positive polynomials first obtained in [245].

Definition 4.3 We denote by $\Gamma_{\text{sub},k}$ the set of all finite-valued submodular cost functions of arity at most k on a Boolean domain D, and we set $\Gamma_{\text{sub}} \overset{\text{def}}{=} \bigcup_k \Gamma_{\text{sub},k}$.

4.4 Reduction to (s, t)-MIN-CUT

In this section, we show equivalence between the minimisation problem for quadratic submodular polynomials and the (s, t)-MIN-CUT problem.

Theorem 4.1 ([147]) (s, t)-MIN-CUT *and the minimisation problem of polynomials over* $\Gamma_{\text{sub},2}$ *are polynomial-time equivalent.*

Proof First we show that any instance $\langle G = \langle V, E \rangle, w, s, t \rangle$, where $w : E \to \mathbb{Q}_{\geq 0}$ and $s, t \in V$, of (s, t)-MIN-CUT is reducible to the minimisation problem for quadratic submodular polynomials:

- every vertex from V is represented by a single Boolean variable;
- by adding a linear term Ms, for large M, we impose a unary constraint on s to take the value 0;
- similarly, by adding a linear term $M(1 - t)$, we impose a unary constraint on t to take the value 1;
- for every edge $\langle u, v \rangle \in E$, we add a submodular quadratic term $av - auv$, where $a = w(\langle u, v \rangle)$ is the weight of the edge $\langle u, v \rangle$ in G (note that $av - auv$ returns a if, and only if, $u = 0$ and $v = 1$, and 0 otherwise).

There is equivalence between (s, t)-cuts in G, that is, subsets of vertices including s but excluding t, and assignments of 0s and 1s to variables in the corresponding polynomial that set s to 0 and t to 1. (Value 0 corresponds to vertices in the cut.)

On the other hand, we show now that any submodular quadratic polynomial can be minimised by a reduction to (s, t)-MIN-CUT.

Let p be an arbitrary submodular quadratic polynomial, that is,

$$p(x_1, \ldots, x_n) = a_0 + \sum_{i=1}^{n} a_i x_i + \sum_{1 \leq i < j \leq n} a_{ij} x_i x_j,$$

where $a_{ij} \leq 0$ for all $1 \leq i < j \leq n$. Then,

$$p = a_0' + \sum_{i \in P} a_i' x_i + \sum_{j \in N} a_j'(1 - x_j) + \sum_{1 \leq i < j \leq n} a_{ij}'(1 - x_i)x_j,$$

where $P \cap N = \emptyset$, $P \cup N = \{1, 2, \ldots, n\}$, $a_{ij}' = -a_{ji}$, and $a_i', a_j', a_{ij}' \geq 0$. (This is known as a *posiform* [34, 84].)

Now p can be easily minimised by a reduction to (s, t)-Min-Cut:

1. vertices of the graph are $x_1, \ldots, x_n$ and two extra vertices s and t;
2. there is an edge going from x_i to x_j with weight a_{ij}';
3. for every $i \in P$, there is an edge going from s to x_i with weight a_i';
4. for every $j \in N$, there is an edge going from x_j to t with weight a_j'.

Again, there is equivalence between (s, t)-cuts in the constructed graph and assignments (which set s to 0 and t to 1) of 0s and 1s to the posiform representation of p. $\qquad\square$

Corollary 4.1 *A quadratic submodular polynomial in n Boolean variables can be minimised in $O(n^3)$ time.*

Proof By Theorem 4.1, and using some standard cubic-time algorithm for (s, t)-Min-Cut [131]. $\qquad\square$

Example 4.2 Consider the following quadratic submodular polynomial:

$$p = 8 + 12x_1 + 7x_2 + 11x_3 - 5x_4 - 7x_5$$
$$- x_1x_2 - 7x_1x_4 - 3x_2x_3 - 4x_3x_4 - 5x_3x_5 - x_4x_5.$$

We can rewrite p as in the proof of Theorem 4.1 as follows:

$$p = 8 + 12x_1 + 7x_2 + 11x_3 - 5x_4 - 7x_5$$
$$+ (1 - x_1)x_2 - x_2 + 7(1 - x_1)x_4 - 7x_4 + 3(1 - x_2)x_3 - 3x_3$$
$$+ 4(1 - x_3)x_4 - 4x_4 + 5(1 - x_3)x_5 - 5x_5 + (1 - x_4)x_5 - x_5$$
$$= -21 + 12x_1 + 6x_2 + 8x_3 + 16(1 - x_4) + 13(1 - x_5)$$
$$+ (1 - x_1)x_2 + 7(1 - x_1)x_4 + 3(1 - x_2)x_3$$
$$+ 4(1 - x_3)x_4 + 5(1 - x_3)x_5 + (1 - x_4)x_5.$$

We can now build a graph G with five vertices corresponding to variables x_1 through x_5 and two extra vertices s and t and add edges accordingly (see Fig. 4.1).

For every assignment v of values 0 and 1 to variables x_1, x_2, x_3, x_4, x_5, $p(v(x_1), \ldots, v(x_5))$ is equal to the size of the (s, t)-cut in G given by v minus 21 (for the constant term in the posiform representation of p). The minimum cut in G, with value 16, is the set $\{s, x_1, x_2, x_3\}$. Therefore, the assignment

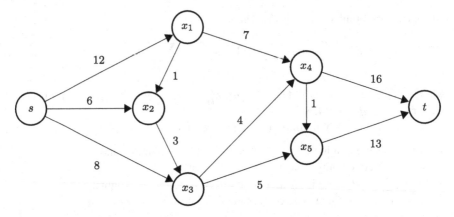

Fig. 4.1 Graph G corresponding to polynomial p (Example 4.2)

$(x_1 \mapsto 0, x_2 \mapsto 0, x_3 \mapsto 0, x_4 \mapsto 0, x_5 \mapsto 0)$ minimises the polynomial p with total value $16 - 21 = -5$.

4.5 Known Classes of Expressible Functions

In this section, we present new and simpler proofs for some known expressibility results. We will focus on three classes of cost function on a Boolean domain: submodular cost functions whose corresponding polynomials have negative coefficients for all terms of degree ≥ 2; $\{0, 1\}$-valued submodular cost functions; and ternary submodular cost functions. We show that cost functions from these three classes are expressible over $\Gamma_{\text{sub},2}$.

Definition 4.4 Define $\Gamma_{\text{neg},k}$ to be the set of all cost functions over a Boolean domain, of arity at most k, whose corresponding polynomials have negative coefficients for all terms of degree greater than or equal to 2. Set $\Gamma_{\text{neg}} \overset{\text{def}}{=} \bigcup_k \Gamma_{\text{neg},k}$.

It is well known that these cost functions, sometimes called *negative-positive*, are submodular [34]. The minimisation of cost functions chosen from Γ_{neg} using min-cuts was first studied in [245], but we give a simplified proof.

Theorem 4.2 ([245]) $\Gamma_{\text{neg}} \subseteq \langle \Gamma_{\text{sub},2} \rangle$.

Proof We use the gadget we have already seen in Example 4.1: Given any polynomial p representing a cost function from Γ_{neg}, we can replace each term $-ax_{i_1}\ldots x_{i_k}$ of p of degree $k \geq 3$, where $a > 0$, by

$$-ax_{i_1}\ldots x_{i_k} = \min_{y \in \{0,1\}} \left\{ a\left(-y + y \sum_{j \in \{1,\ldots,k\}} (1 - x_{i_j})\right) \right\}.$$

In this way, we introduce a new variable for every term of degree ≥ 3. $\square$

Corollary 4.2 *A polynomial p in n Boolean variables over Γ_{neg} can be minimised in $O((n+r)^3)$ time, where r is the number of terms of degree 3 or greater in p.*

Next we consider the class of cost functions over a Boolean domain which take only the cost values 0 and 1. (Such constraints can be used to model optimisation problems such as MAX-CSP; see [59].)

Definition 4.5 Define $\Gamma_{\{0,1\},k}$ to be the set of all $\{0,1\}$-valued submodular cost functions over a Boolean domain, of arity at most k, and set $\Gamma_{\{0,1\}} \stackrel{\text{def}}{=} \bigcup_k \Gamma_{\{0,1\},k}$.

The minimisation of submodular cost functions from $\Gamma_{\{0,1\}}$ has been studied in [87], where they have been called *2-monotone* functions. The equivalence of 2-monotone and submodular cost functions and a generalisation of 2-monotone functions to non-Boolean domains[3] has been shown in [59].

Definition 4.6 A cost function ϕ is called *2-monotone* if there exist two sets $A, B \subseteq \{1, \ldots, n\}$ such that $\phi(\mathbf{x}) = 0$ if $A \subseteq \mathbf{x}$ or $\mathbf{x} \subseteq B$ and $\phi(\mathbf{x}) = 1$ otherwise (where $A \subseteq \mathbf{x}$ means $\forall i \in A, x_i = 1$ and $\mathbf{x} \subseteq B$ means $\forall i \notin B, x_i = 0$).

Using the notion of expressive power we are able to give a very simple proof of the following result:

Theorem 4.3 ([87]) $\Gamma_{\{0,1\}} \subseteq \langle \Gamma_{\mathrm{sub},2} \rangle$.

Proof Any 2-monotone cost function ϕ can be expressed over $\Gamma_{\mathrm{sub},2}$ using two extra variables, y_1, y_2:

$$\phi(\mathbf{x}) = \min_{y_1, y_2 \in \{0,1\}} \left\{ (1 - y_1)y_2 + y_1 \sum_{i \in A}(1 - x_i) + (1 - y_2)\sum_{i \notin B} x_i \right\}. \qquad \square$$

Corollary 4.3 *A polynomial p in n Boolean variables over $\Gamma_{\{0,1\}}$ can be minimised in $O((n+r)^3)$ time, where r is the number of 2-monotone cost functions represented in p.*

Finally, we consider the class $\Gamma_{\mathrm{sub},3}$ of ternary submodular cost functions over a Boolean domain. This class has been studied in [20], from where we obtain the following useful characterisation of cubic submodular polynomials.

---

[3]In [59], 2-monotone cost functions are defined over lattice-ordered sets and called *generalised 2-monotone* cost functions. In [191], these are called just *2-monotone* cost functions.

Lemma 4.2 ([20]) *A cubic polynomial $p(x_1, \ldots, x_n)$ over Boolean variables represents a submodular cost function if, and only if, it can be written as*

$$p(x_1, \ldots, x_n) = a_0 + \sum_{\{i\} \in C_1^+} a_i x_i - \sum_{\{i\} \in C_1^-} a_i x_i - \sum_{\{i,j\} \in C_2} a_{ij} x_i x_j$$

$$+ \sum_{\{i,j,k\} \in C_3^+} a_{ijk} x_i x_j x_k - \sum_{\{i,j,k\} \in C_3^-} a_{ijk} x_i x_j x_k,$$

where C_2 denotes the set of quadratic terms, and C_i^+ (C_i^-) denotes the set of terms of degree i with positive (negative) coefficients, for $i = 1, 3$, and

1. $a_i, a_{ij}, a_{ijk} \geq 0$ ($\{i\} \in C_1^+ \cup C_1^-, \{i, j\} \in C_2, \{i, j, k\} \in C_3^+ \cup C_3^-$), *and*
2. $\forall \{i, j\} \in C_2, a_{ij} + \sum_{k \mid \{i,j,k\} \in C_3^+} a_{ijk} \leq 0$.

This characterisation, together with the notion of expressive power, allows us to obtain a very simple proof of the following result:

Theorem 4.4 ([20]) $\Gamma_{\mathsf{sub},3} \subseteq \langle \Gamma_{\mathsf{sub},2} \rangle$.

Proof Let p be a polynomial representing an arbitrary cost function in $\Gamma_{\mathsf{sub},3}$. By submodularity, all quadratic terms in p are nonpositive. We already know how to express a negative cubic term using a gadget over $\Gamma_{\mathsf{sub},2}$ (Theorem 4.2). To express a positive cubic term, consider the following identity:

$$x_i x_j x_k - x_i x_j - x_i x_k - x_j x_k = \min_{y \in \{0,1\}} \left\{ (1 - x_i - x_j - x_k) y \right\}.$$

Hence, we can replace a positive cubic term $a_{ijk} x_i x_j x_k$ in p with

$$\min_{y \in \{0,1\}} \left\{ a_{ijk} (1 - x_i - x_j - x_k) y + a_{ijk} (x_i x_j + x_i x_k + x_j x_k) \right\}.$$

It remains to check that all quadratic coefficients of the resulting polynomial are nonpositive. However, this is ensured by the second condition from Lemma 4.2. $\square$

Corollary 4.4 *A cubic submodular polynomial p in n Boolean variables can be minimised in $O((n + r)^3)$, where r is the number of cubic terms in p.*

Remark 4.6 The proof of Theorem 4.4 given in [20] was obtained using a different approach based on the so-called conflict graphs of a supermodular polynomial (see [34, 84]). Such graphs have been shown to be bipartite, and therefore the problem of finding a maximum weight stable set can be reduced to a flow problem. However, the resulting time complexity is the same.

4.6 New Classes of Expressible Functions

In this section, we present new classes of submodular cost functions that can be expressed by binary submodular cost functions. First, we derive a necessary condition for a 4-ary cost function over a Boolean domain to be submodular. We also present some sufficient conditions, which give rise to new classes of submodular cost functions that can be expressed over $\Gamma_{\text{sub},2}$, and hence be minimised efficiently. We prove the sufficient conditions first for 4-ary submodular cost functions and then generalise them to k-ary submodular cost functions for every $k \geq 4$.

We start with submodular cost functions of arity 4. One might hope to obtain a simple characterisation of 4-ary submodular cost functions over a Boolean domain similar to that of Lemma 4.2. However, it has been shown that testing whether a given quartic Boolean polynomial is submodular is co-NP-complete [83, 124]. Hence, one is unlikely to find a polynomial-time checkable characterisation, as this would prove that P = NP. However, we obtain the following necessary condition.

Lemma 4.3 *If a quartic polynomial $p(x_1, \ldots, x_n)$ over Boolean variables represents a submodular cost function, then it can be written such that, for all $\{i, j\} \in C_2$:*

1. $a_{ij} \leq 0$, *and*
2. $a_{ij} + \sum_{k | \{i,j,k\} \in C_3^+} a_{ijk} + \sum_{k,l | \{i,j,k,l\} \in C_4^+} a_{ijkl} + F_{ij} \leq 0$, *where*

$$F_{ij} = \sum_{k|\{i,j,k\}\in C_3^- \wedge \{i,j,k,.\}\in C_4^+} a_{ijk} + \sum_{k,l|\{i,j,k,l\}\in C_4^- \wedge \{i,j,k,.\},\{i,j,l,.\}\in C_4^+} a_{ijkl},$$

C_2 *denotes the set of quadratic terms, and C_i^+ (C_i^-) denotes the set of terms of degree i with positive (negative) coefficients, for $i = 3, 4$.*

Proof Let p be a quartic submodular polynomial and let i and j be given; then $\delta_{ij}(x_1, \ldots, x_{i-1}, x_{i+1}, \ldots, x_{j-1}, x_{j+1}, \ldots, x_n)$, the second-order derivative of p with respect to the ith and jth variable, is equal to

$$\delta_{ij} = a_{ij} + \sum_{k|\{i,j,k\}\in C_3^+} a_{ijk}x_k + \sum_{k,l|\{i,j,k,l\}\in C_4^+} a_{ijkl}x_kx_l$$

$$- \sum_{k|\{i,j,k\}\in C_3^-} a_{ijk}x_k - \sum_{k,l|\{i,j,k,l\}\in C_4^-} a_{ijkl}x_kx_l.$$

Consider an assignment which sets $x_k = 1$ if $k = i$ or $k = j$, and $x_k = 0$ otherwise. From submodularity, $a_{ij} \leq 0$, which proves the first condition. By setting $x_k = 1$ for all k such that $\{i, j, k\} \in C_3^+$, and $x_k = x_l = 1$ for all k, l such that $\{i, j, k, l\} \in C_4^+$, we get the second condition. We set to 1 all variables that occur in some positive cubic or quartic term. The second condition then says that the sum of all these positive coefficients minus those that are forced, by our setting of variables, to be 1 (F_{ij}) is at most 0. (Note that this also proves Lemma 4.2.)   □

Next we show a useful example of a 4-ary submodular cost function that can be expressed over the binary submodular cost functions using one extra variable.

Example 4.3 Let ϕ be the 4-ary cost function defined as follows: $\phi(\mathbf{x}) = \min\{2k, 5\}$, where k is the number of 0s in $\mathbf{x} \in \{0, 1\}^4$. The corresponding quartic polynomial representing ϕ is

$$p(x_1, x_2, x_3, x_4) = 5 + x_1 x_2 x_3 x_4 - x_1 x_2 - x_1 x_3 - x_1 x_4 - x_2 x_3 - x_2 x_4 - x_3 x_4.$$

By considering second-order derivatives of p, it can be checked that p is submodular. For instance, $\delta_{12}(x_3, x_4)$, the second-order derivative of p with respect to the first two variables, is equal to $x_1 x_2 x_3 x_4 - 1$. Clearly, $\delta_{12}(x_3, x_4) \leq 0$. It can be shown by a simple case analysis that p cannot be expressed as a quadratic polynomial with nonpositive quadratic coefficients (from the definition of p, the polynomial would have to be $5 - x_1 x_2 - x_1 x_3 - x_1 x_4 - x_2 x_3 - x_2 x_4 - x_3 x_4$, which is not equal to p on $x_1 = x_2 = x_3 = x_4 = 1$).

However, p can be expressed over $\Gamma_{\text{sub},2}$ using just one extra variable, via the following gadget:

$$p(x_1, x_2, x_3, x_4) = \min_{y \in \{0,1\}} \left\{ 5 + (3 - 2x_1 - 2x_2 - 2x_3 - 2x_4)y \right\}.$$

Definition 4.7 Using the same notation as in Lemma 4.3, define $\Gamma_{\text{suff},4}$ to be the set of all 4-ary submodular cost functions over a Boolean domain whose corresponding quartic polynomials satisfy, for every $i < j$,

$$a_{ij} + \sum_{k | \{i,j,k\} \in C_3^+} a_{ijk} + \sum_{k,l | \{i,j,k,l\} \in C_4^+} a_{ijkl} \leq 0. \tag{4.1}$$

Theorem 4.5 $\Gamma_{\text{suff},4} \subseteq \langle \Gamma_{\text{sub},2} \rangle$.

Proof Let $\phi \in \Gamma_{\text{suff},4}$ and let p be the corresponding polynomial which represents ϕ. First, replace all negative cubic and quartic terms using the construction in Theorem 4.2. As in the proof of Theorem 4.4, replace every positive cubic term $a_{ijk} x_i x_j x_k$ in p with

$$\min_{y \in \{0,1\}} \left\{ a_{ijk}(1 - x_i - x_j - x_k)y + a_{ijk}(x_i x_j + x_i x_k + x_j x_k) \right\}.$$

Using the same construction as in Example 4.3, replace every positive quartic term $a_{ijkl} x_i x_j x_k x_l$ with

$$\min_{y \in \{0,1\}} \left\{ a_{ijkl}(3 - 2x_i - 2x_j - 2x_k - 2x_l)y \right.$$
$$\left. + a_{ijkl}(x_i x_j + x_i x_k + x_i x_l + x_j x_k + x_j x_l + x_k x_l) \right\}.$$

It only remains to check that all quadratic coefficients in the resulting polynomial are nonpositive. However, this is ensured by the definition of $\Gamma_{\text{suff},4}$ and by the choice of the gadgets.   □

Our next example shows that $\Gamma_{\text{suff},4} \subsetneq \Gamma_{\text{sub},4}$.

Example 4.4 Define a 4-ary submodular cost function ϕ as follows: $\phi(\mathbf{x}) = \min(3k, 7) + 2y + z$, where k is the number of 0s in $\mathbf{x} \in \{0, 1\}^4$; $y = 1$ if, and only if, $\mathbf{x} = \langle 1, 1, 1, 0 \rangle$ (and 0 otherwise), and $z = 1$ if, and only if, $\mathbf{x} = \langle 1, 1, 0, 0 \rangle$ (and 0 otherwise). The corresponding polynomial representing ϕ is

$$p(x_1, x_2, x_3, x_4) = 7 + 2x_1x_2x_3x_4 - 2x_1x_2x_4 - x_1x_3x_4 - x_2x_3x_4$$
$$- x_1x_3 - x_1x_4 - x_2x_3 - x_2x_4 - x_3x_4.$$

By considering the second-order derivatives of p, it can easily be checked that ϕ is submodular. However, $\phi \notin \Gamma_{\text{suff},4}$: for $i = 1$ and $j = 2$, the expression in Eq. (4.1) gives 2. Hence Theorem 4.5 does not apply to ϕ.

As in Example 4.3, by a simple case analysis (system of equations), it can be shown that ϕ cannot be expressed over $\Gamma_{\text{sub},2}$ without extra variables or with just one extra variable. However, the following gadget shows that ϕ is in fact expressible over $\Gamma_{\text{sub},2}$ using just two extra variables:

$$p(x_1, x_2, x_3, x_4) = 7 - x_1x_4 - x_2x_4 - x_3x_4$$
$$+ \min_{y_1, y_2 \in \{0,1\}} \{2y_1 + 3y_2 - y_1y_2 - y_1(x_1 + x_2 + 2x_3)$$
$$- y_2(x_1 + x_2 + 2x_4)\}.$$

We now generalise the expressibility result of submodular cost functions from $\Gamma_{\text{suff},k}$ to subclasses of submodular cost functions of arbitrary arities.

Definition 4.8   We define $\Gamma_{\text{suff},k}$ to be the set of all k-ary submodular cost functions over a Boolean domain whose corresponding polynomials satisfy, for every $1 \le i < j \le k$,

$$a_{ij} + \sum_{s=1}^{k-2} \sum_{\{i,j,i_1,\ldots,i_s\} \in C_{s+2}^+} a_{i,j,i_1,\ldots,i_s} \le 0,$$

where C_i^+ denotes the set of terms of degree i with positive coefficients.

In other words, for any $1 \le i < j \le k$, the sum of a_{ij} and all positive coefficients of cubic and higher-degree terms that include x_i and x_j is nonpositive.

Theorem 4.6 *For every $k \ge 4$, $\Gamma_{\text{suff},k} \subseteq \langle \Gamma_{\text{sub},2} \rangle$.*

Proof First we show how to uniformly generate gadgets over $\Gamma_{\mathrm{sub},2}$ for polynomials of the following type:

$$p_k(x_1, \ldots, x_k) = \prod_{i=1}^{k} x_i - \sum_{1 \le i < j \le k} x_i x_j.$$

Note that $p_k(\mathbf{x}) = -\binom{m}{2}$, where m is the number of 1s in $\mathbf{x}$, and $\binom{0}{2} = \binom{1}{2} = 0$, unless $m = k$ ($\mathbf{x}$ consists of 1s only), in which case $p_k(\mathbf{x}) = -\binom{m}{2} + 1$.

We claim, that for any $k \ge 4$, the following, denoted by $\mathcal{P}_k$, is a gadget for p_k:

$$p_k(x_1, \ldots, x_k) = \min_{y_0, \ldots, y_{k-4} \in \{0,1\}} \left\{ y_0 \left(3 - 2 \sum_{i=1}^{k} x_i \right) + \sum_{j=1}^{k-4} y_j \left(2 + j - \sum_{i=1}^{k} x_i \right) \right\}.$$

Notice that in the case of $k = 4$, the gadget corresponds to the gadget used in the proof of Theorem 4.5, and therefore the base case is proved. We proceed by induction on k. Assume that $\mathcal{P}_i$ is a gadget for p_i for every $i \le k$. We prove that $\mathcal{P}_{k+1}$ is a gadget for p_{k+1}.

Firstly, take the gadget $\mathcal{P}_k$ for p_k, and replace every sum $\sum_{i=1}^{k} x_i$ with $\sum_{i=1}^{k+1} x_i$. We denote the new gadget $\mathcal{P}'$. By the inductive hypothesis, it is not difficult to see that $\mathcal{P}'$ is a valid gadget for p_{k+1} on all assignments with at most $k - 1$ 1s. Also, on any assignment with exactly k 1s, $\mathcal{P}'$ returns $-\binom{k}{2} + 1$. On the assignment with $k + 1$ 1s, $\mathcal{P}'$ returns $-\binom{k}{2} + 1 - 2 - 1(k - 4)$. This can be simplified as follows:

$$-\binom{k}{2} + 1 - 2 - k + 4 = -\binom{k}{2} + 1 - k + 2$$

$$= -\left(\binom{k}{2} + \binom{k}{1} \right) + 1 + 2$$

$$= -\binom{k+1}{2} + 1 + 2.$$

Hence $\mathcal{P}'$ is *almost* a gadget for p_{k+1}: we only need to subtract 1 on an assignment that has exactly k 1s, and subtract 2 on the assignment consisting of 1s only. But this is exactly what $\min_{y_{k-3} \in \{0,1\}} \{y_{k-3}(2 + (k - 3) - \sum_{i=1}^{k+1} x_i)\}$ does. Therefore, we have established that $\mathcal{P}_{k+1}$ is a gadget for p_{k+1} over $\Gamma_{\mathrm{sub},2}$ with $k - 3$ extra variables.

Given a cost function $\phi \in \Gamma_{\mathrm{suff},k}$, let p be the corresponding polynomial that represents ϕ. By the construction in Theorem 4.2, we can replace all negative terms of degree ≥ 3. By the constructions in Theorems 4.4 and 4.5, we can replace all positive cubic and quartic terms. Now for any positive term of degree d, $5 \le d \le k$, we replace the term with the gadget $\mathcal{P}_d$ and add $\sum_{1 \le i < j \le k} x_i x_j$ back in. This construction works if all quadratic coefficients of the resulting polynomial are nonpositive. However, this is ensured by the definition of $\Gamma_{\mathrm{suff},k}$ and by the choice of the gadgets. $\qquad\square$

Corollary 4.5 *A polynomial p in n Boolean variables over Γ_{suff} can be minimised in $O((n + r)^3)$ time, where r is the number of cost functions of arity 3 or greater represented in p.*

4.7 Summary

We have studied the expressive power of binary Boolean submodular functions. In the pseudo-Boolean optimisation language, we studied classes of submodular functions whose polynomial representation can be decomposed into a sum of binary submodular polynomials, over a possibly larger set of variables.

We have shown new and simpler proofs of known results for a range of special classes. Moreover, we have shown a new class of submodular functions of arbitrary arities that can be expressed by binary submodular functions. Consequently, we have shown in a uniform way that all these classes of submodular functions can be minimised efficiently by being reduced to the (s, t)-MIN-CUT problem. The same new class of functions has been obtained independently by Zalesky [271] using different gadgets.

Chapter 5 studies the question of which submodular functions are expressible by binary submodular functions in more depth, obtains another general class of expressible functions, and discusses applications of these results to VCSP instances with submodular constraints.

4.7.1 Related Work

There has been a lot of research on subclasses of pseudo-Boolean functions that can be minimised[4] using (s, t)-MIN-CUT or reduced to this problem by switching a subset of variables (see [34] for a survey). Note that switching a subset of variables does not preserve submodularity.

In Sect. 4.5, we discussed the class of *negative-positive* (also known as *almost-positive*) functions. This class has been studied in [8, 232, 233, 245]. Functions which can be made negative-positive by switching a subset of variables are called *unate* functions. The class of unate functions has been studied and shown to be recognisable in polynomial time in [259]. Note that unate functions are not in general submodular.

The class of cost function that can be made submodular by permuting the domains of all variables is called *renamable, permutable,* or *switchable* submodular. Testing whether a given binary VCSP instance consists of renamable submodular cost functions can be done in polynomial time for any finite domain [254].

---

[4]Most papers on pseudo-Boolean optimisation deal with the maximisation problem, and therefore talk about supermodular functions, rather than submodular functions. However, the maximisation problem of supermodular functions is equivalent to the minimisation problem of submodular functions.

4.7.2 Open Problems

Are there other subclasses of submodular functions that can be minimised more efficiently than all submodular functions? (One such example, which is different from the classes considered in this chapter, is the class of symmetric submodular functions [238].)

Chapter 5
Non-expressibility of Submodular Languages

A man should look for what is, and not for what he thinks should be.
Albert Einstein

5.1 Introduction

In Chap. 4, we studied the expressive power of binary submodular functions. We have seen that showing that a class C of submodular functions is expressible by binary submodular functions is equivalent to showing that functions from C can be minimised via (s, t)-MIN-CUT. Furthermore, showing that a class C of submodular functions is *not* expressible by binary submodular functions is equivalent to showing that the SFM$_b$ problem for functions from C *cannot* be reduced to the (s, t)-MIN-CUT problem via the expressibility reduction.

It has previously been an open problem whether all Boolean submodular functions can be decomposed into a sum of binary submodular functions over a possibly larger set of variables. This problem has been considered within several different contexts in computer science, including computer vision, artificial intelligence, and pseudo-Boolean optimisation. Using the connection between the expressive power of valued constraints and certain algebraic properties of functions described in earlier chapters, we answer this question negatively. This chapter is based on [275].

5.2 Results

The series of positive expressibility results in Chap. 4 naturally raises the following question:

Problem 5.1 Are *all* submodular polynomials expressible by binary submodular polynomials, over a possibly larger set of variables?

This chapter is reprinted from *Discrete Applied Mathematics*, **157**(15), S. Živný, D.A. Cohen, and P.G. Jeavons, The Expressive Power of Binary Submodular Functions, 3347–3358, Copyright (2009), with permission from Elsevier.

Each of the above expressibility results was obtained by an ad hoc construction, and no general technique[1] has previously been proposed that is sufficiently powerful to address Problem 5.1.

Using the algebraic approach to characterising the expressive power of valued constraints developed by Cohen et al. in [65, 68] and discussed in Chap. 2, we are able to give a negative answer to Problem 5.1: we show that there exist submodular polynomials of arity 4 that cannot be expressed by binary submodular polynomials. More precisely, we characterise exactly which submodular polynomials of arity 4 are expressible by binary submodular polynomials and which are not. In addition, we show that any submodular polynomial of arity 4 is either expressible by binary submodular polynomials with only a small number of extra variables, or is not expressible at all.

On the way to establishing these results we show that two broad families of submodular functions, known as *upper fans* and *lower fans*, are all expressible by binary submodular functions. This provides a large new class of submodular polynomials of all arities that are expressible by binary submodular polynomials and hence are solvable efficiently by reduction to (s, t)-MIN-CUT. We use the expressibility of this family, and the existence of non-expressible functions, to refute a conjecture from [236] on the structure of the extreme rays of the cone of Boolean submodular functions, and suggest a more refined conjecture of our own.

The rest of this chapter is organised as follows. In Sect. 5.3, we show some new classes of submodular functions to be expressible by binary submodular functions. In particular, we show that all upper fans and lower fans of arbitrary arities are expressible by binary submodular functions. In Sect. 5.4, we characterise precisely the multimorphisms of all binary submodular functions. Moreover, we characterise the fractional clone of all binary submodular functions. The characterisation of the multimorphisms of binary submodular functions helps us to prove, in Sect. 5.5, the main result of this chapter: there are submodular functions that are *not* expressible by binary submodular functions. Moreover, we characterise precisely which submodular functions of arity 4 can be expressed by binary submodular functions. Our results imply limitations on the expressibility reduction and establish for the first time that it cannot be used in general to minimise arbitrary submodular functions. Finally, we refute a conjecture of Promislow and Young [236] on the structure of the extreme rays of the cone of Boolean submodular functions. In Sect. 5.6, we present some results on the recognition problem for submodular functions. Section 5.7 summarises this chapter and comments on related work.

5.3 Expressibility of Upper Fans and Lower Fans

We begin by defining some particular families of submodular cost functions, first described in [236], that will turn out to play a central role in our analysis.

---

[1]For example, standard combinatorial counting techniques cannot resolve this question because we allow arbitrary real-valued coefficients in submodular polynomials. We also allow an arbitrary number of additional variables.

Definition 5.1 Let L be a lattice. We define the following cost functions on L:

- For any set A of pairwise incomparable elements $\{a_1, \ldots, a_m\} \subseteq L$, such that each pair of distinct elements (a_i, a_j) has the same least upper bound, $\bigvee A$, the following cost function is called an *upper fan*:

$$\phi_A(x) \stackrel{\text{def}}{=} \begin{cases} -2 & \text{if } x \geq \bigvee A, \\ -1 & \text{if } x \not\geq \bigvee A, \text{ but } x \geq a_i \text{ for some } i, \\ 0 & \text{otherwise.} \end{cases}$$

- For any set B of pairwise incomparable elements $\{a_1, \ldots, a_m\} \subseteq L$, such that each pair of distinct elements (a_i, a_j) has the same greatest lower bound, $\bigwedge B$, the following cost function is called a *lower fan*:

$$\phi_B(x) \stackrel{\text{def}}{=} \begin{cases} -2 & \text{if } x \leq \bigwedge B, \\ -1 & \text{if } x \not\leq \bigwedge B, \text{ but } x \leq a_i \text{ for some } i, \\ 0 & \text{otherwise.} \end{cases}$$

We call a cost function a *fan* if it is either an upper fan or a lower fan. Note that our definition of fans is slightly more general than the definition in [236]. In particular, we allow the set A to be empty, in which case the corresponding upper fan ϕ_A is a constant function. It is not hard to show that all fans are submodular [236].

Note that when $D = \{0, 1\}$, the set D^n with the product ordering is isomorphic to the lattice of all subsets of an n-element set ordered by inclusion. Hence, a cost function on a Boolean domain can be viewed as a cost function defined on a lattice of subsets, and we can apply Definition 5.1 to identify certain Boolean functions as upper fans or lower fans, as the following example indicates.

Example 5.1 Let $A = \{I_1, \ldots, I_r\}$ be a set of subsets of $\{1, 2, \ldots, n\}$ such that for all $i \neq j$ we have $I_i \not\subseteq I_j$ and $I_i \cup I_j = \bigcup A$.

By Definition 5.1, the corresponding upper fan function ϕ_A has the following polynomial representation:

$$p(x_1, \ldots, x_n) = (r-2) \prod_{i \in \bigcup A} x_i - \prod_{i \in I_1} x_i - \cdots - \prod_{i \in I_r} x_i.$$

Remark 5.1 Any permutation of a set D gives rise to an automorphism of cost functions over D. In particular, for any cost function f on a Boolean domain D, the *dual* of f is the corresponding cost function that results from exchanging the values 0 and 1 for all variables. In other words, if p is the polynomial representation of f, then the dual of f is the cost function whose polynomial representation is obtained from p by replacing all variables x with $1 - x$. Observe that, due to symmetry, taking the dual preserves submodularity and expressibility by binary submodular cost functions.

It follows from Definition 5.1 that upper fans are duals of lower fans, and vice versa.

Definition 5.2 We denote by $\Gamma_{\text{fans},k}$ the set of all fans of arity at most k on a Boolean domain D, and we set $\Gamma_{\text{fans}} \overset{\text{def}}{=} \bigcup_k \Gamma_{\text{fans},k}$.

Our next result shows that $\Gamma_{\text{fans}} \subseteq \langle \Gamma_{\text{sub},2} \rangle$; that is, fans of all arities are expressible by binary submodular functions.

Theorem 5.1 *Any fan on a Boolean domain D is expressible by binary submodular functions on D using at most $1 + \lfloor m/2 \rfloor$ extra variables, where m is the degree of its polynomial representation.*

Proof Since upper fans are dual to lower fans, it is sufficient to establish the result for upper fans only.

Let $A = \{I_1, \ldots, I_r\}$ be a set of subsets of $\{1, 2, \ldots, n\}$ such that for all $i \neq j$ we have $I_i \not\subseteq I_j$ and $I_i \cup I_j = \bigcup A$, and let ϕ_A be the corresponding upper fan, as specified by Definition 5.1. The polynomial representation of ϕ_A, $p(x_1, \ldots, x_n)$, is given in Example 5.1.

The degree of p is equal to the total number of variables occurring in it, which will be denoted by m. Note that $m = |\bigcup A|$.

If $r = 0$, then ϕ_A is constant, so the result holds trivially. If $r = 1$, we have $A = \{I\}$, where $I = \{i_1, \ldots, i_m\}$ and the polynomial representation of ϕ_A is $-2x_{i_1} x_{i_2} \cdots x_{i_m}$. In this case, it was shown in Example 4.1 that ϕ_A can be expressed by quadratic functions using one extra variable, as follows:

$$-2x_{i_1} x_{i_2} \cdots x_{i_m} = \min_{y \in \{0,1\}} \left\{ 2y \left((m-1) - \sum_{i \in I} x_i \right) \right\}.$$

For the case when $r > 1$, we first note that any $i \in \bigcup A$ must belong to all the elements of A except at most one (otherwise there would be two elements of A, say I_i and I_j, such that $I_i \cup I_j \neq \bigcup A$, which contradicts the choice of A).

We will say that two elements of $\bigcup A$ are *equivalent* if they occur in exactly the same elements of A; that is, $i_1, i_2 \in \bigcup A$ are equivalent if $i_1 \in I_j \Leftrightarrow i_2 \in I_j$ for all $j \in \{i, \ldots, r\}$. Equivalent elements i_1 and i_2 of $\bigcup A$ can be merged by replacing them with a single new element. In the polynomial representation of ϕ_A this corresponds to replacing the variables x_{i_1} and x_{i_2} with a single new variable, z, corresponding to their product. Note that the number of equivalence classes of size 2 or greater is at most $\lfloor m/2 \rfloor$.

After completing all such merging, we obtain a new set $A' = \{I_1', \ldots, I_{r'}'\}$ with the property that $|I_i'| = m' - 1$ for every i, where $m' = |\bigcup A'|$ is the size of the common join of any $I_i', I_j' \in A'$. This set has a corresponding new upper fan, $\phi_{A'}$, over the new merged variables.

To complete the proof we will construct a simple gadget for expressing $\phi_{A'}$, and show how to use this to obtain a gadget for expressing the original upper fan ϕ_A.

Note that the sets I_i' are subsets of $\bigcup A'$, each of size $m' - 1$. Any such subset is uniquely determined by its single missing element. We denote by K the set of elements occurring in *all* sets I_i' and by L the set of elements that are missing from one of these subsets. Clearly, $|K| + |L| = m'$. We claim that the following polynomial is a gadget for expressing ϕ'_A:

$$p'(z_1, \ldots, z_{m'}) = \min_{y \in \{0,1\}} \left\{ y \left(2(m'-1) - |L| - \sum_{i \in L} z_i - 2 \sum_{i \in K} z_i \right) \right\}.$$

To establish this claim, we will compute the value of p', for each possible assignment to the variables $z_1, \ldots, z_{m'}$. Denote by k_0 the number of 0s assigned to variables in K, and by l_0 the number of 0s assigned to variables in L. Then we have:

$$\begin{aligned}
p'(z_1, \ldots, z_{m'}) &= \min_{y \in \{0,1\}} y \left(2m' - 2 - |L| - \sum_{i \in L} z_i - 2 \sum_{i \in K} z_i \right) \\
&= \min_{y \in \{0,1\}} y \left(2m' - 2 - |L| - (|L| - l_0) - 2(m' - |L| - k_0) \right) \\
&= \min_{y \in \{0,1\}} y \left(2m' - 2 - 2|L| + l_0 - 2m' + 2|L| + 2k_0 \right) \\
&= \min_{y \in \{0,1\}} y(-2 + 2k_0 + l_0).
\end{aligned}$$

Hence if $k_0 = l_0 = 0$, then p' takes the value -2. If $k_0 = 0$ and $l_0 = 1$, then p' takes the value -1. In all other cases (that is, $k_0 > 0$ or $l_0 > 1$), p' takes the value 0. By Definition 5.1, this means that p' is the (unique) polynomial representation for $\phi_{A'}$. Note that p' uses just one extra variable, y.

Finally, we show how to obtain a gadget for the original upper fan ϕ_A from the polynomial p'. Each variable in p' represents an equivalence class of elements of $\bigcup A$, so it can be replaced by a term consisting of the product of the variables in this equivalence class. In this way we obtain a new polynomial over the original variables containing linear and negative quadratic terms together with negative higher-order terms (cubic or above) corresponding to every equivalence class with two or more elements. However, each of these higher-order terms can itself be expressed by a quadratic submodular polynomial, by introducing a single extra variable, as shown in the case when $r = 1$, above. Therefore, combining each of these polynomials, the total number of new variables introduced is at most $1 + \lfloor m/2 \rfloor$. $\qquad \square$

All of the expressibility results from Chap. 4 can be obtained as simple corollaries of Theorem 5.1, as the following examples indicate. In other words, Theorem 5.1 generalises all previously known classes of submodular functions that can be expressed by binary submodular functions.

Example 5.2 Any negative monomial $-x_1 x_2 \cdots x_m$ is a positive multiple of an upper fan, and the positive linear monomial x_1 is equal to $-(1 - x_1) + 1$, so it is a positive multiple of a lower fan, plus a constant. Hence all negative-positive submodular polynomials are contained in $\mathrm{Cone}(\Gamma_{\mathrm{fans}})$, and by Theorem 5.1, they are expressible by binary submodular polynomials, as originally shown in [245], and also in Theorem 4.2.

Example 5.3 Any cubic submodular polynomial can be expressed as a positive sum of upper fans [236]. Hence, by Theorem 5.1, all cubic submodular polynomials are expressible by binary submodular polynomials, as originally shown in [20], and also in Theorem 4.4.

Example 5.4 Recall from Definition 4.6 that a Boolean cost function ϕ is called *2-monotone* [87] if there exist two sets $R, S \subseteq \{1, \ldots, n\}$ such that $\phi(\mathbf{x}) = 0$ if $R \subseteq \mathbf{x}$ or $\mathbf{x} \subseteq S$ and $\phi(\mathbf{x}) = 1$ otherwise (where $R \subseteq \mathbf{x}$ means $\forall i \in R, x[i] = 1$ and $\mathbf{x} \subseteq S$ means $\forall i \notin S, x[i] = 0$). It was shown in [59, Proposition 2.9] that a 2-valued Boolean cost function is 2-monotone if, and only if, it is submodular.

For any 2-monotone cost function defined by the sets of indices R and S, it is straightforward to check that $\phi = \min_{y \in \{0,1\}} y(1 + \phi_A/2) + (1 - y)(1 + \phi_B/2)$, where ϕ_A is the upper fan defined by $A = \{R\}$ and ϕ_B is the lower fan defined by $B = \{\overline{S}\}$. Note that the function $y\phi_A$ is an upper fan, and the function $(1 - y)\phi_B$ is a lower fan. Hence, by Theorem 5.1, all 2-monotone polynomials are expressible by binary submodular polynomials, and solvable by reduction to (s, t)-MIN-CUT, as originally shown in [87], and also in Theorem 4.3.

Example 5.5 A much-studied subclass of submodular functions which can be minimised via (s, t)-MIN-CUT is the class of *polar* (also known as *homogeneous* [20]) functions [83]. Polar functions are functions that have a posiform representation (this means that all coefficients except the constant one are nonnegative) such that all monomials consist of only variables or negated variables [20].

More formally, a polynomial is called polar if it can be expressed as a sum of terms of the form $a x_1 x_2 \cdots x_k$ or $a(1 - x_1)(1 - x_2) \cdots (1 - x_k)$ with positive coefficients a, together with a constant term. It was observed in [20] that all polar polynomials are supermodular. Hence in our context it makes sense to talk about negated polar polynomials, which are required to have all coefficients except the constant one nonpositive. It follows from results in [20] that all negated polar polynomials are submodular.

It is known that for binary and ternary functions, negated polar functions are precisely submodular functions [83]. Moreover, it is known that for functions of arity 4, negated polar functions are strictly included in the class of submodular functions of arity 4 [20].

As every negated term $-a x_1 x_2 \cdots x_k$ is a positive multiple of an upper fan, and every negated term $-a(1 - x_1)(1 - x_2) \cdots (1 - x_k)$ is a positive multiple of a lower fan, by Theorem 5.1, all cost functions that are the negations of polar polynomials are expressible by binary submodular polynomials, and solvable by reduction to (s, t)-MIN-CUT, as originally shown in [20].

However, Theorem 5.1 also provides many new functions of all arities that have not previously been shown to be expressible by binary submodular functions, as the following example indicates.

Example 5.6 The function $2x_1 x_2 x_3 x_4 - x_1 x_2 x_3 - x_1 x_2 x_4 - x_1 x_3 x_4 - x_2 x_3 x_4$ belongs to $\Gamma_{\text{fans},4}$, but does not belong to any class of submodular functions that has previously been shown to be expressible by binary submodular functions. In particular, it does not belong to the class Γ_{new} identified in Chap. 4 (cf. Definition 4.7).

5.4 Characterisation of Mul($\Gamma_{sub,2}$) and fPol($\Gamma_{sub,2}$)

Since we have seen that a cost function can only be expressed by a given set of cost functions if it has the same multimorphisms as them, we now investigate the multimorphisms and fractional polymorphisms of $\Gamma_{sub,2}$.

Notation 5.1 A function $\mathcal{F} : D^k \to D^k$ is called *conservative* if, for each possible choice of $x_1, \ldots, x_k$, the tuple $\mathcal{F}(x_1, \ldots, x_k)$ is a permutation of $x_1, \ldots, x_k$ (though different inputs may be permuted in different ways).

Lemma 5.1 *Let Γ be a valued constraint language including all unary cost functions. Then any multimorphism $\mathcal{F}$ of Γ, $\mathcal{F} \in \mathsf{Mul}(\Gamma)$, is conservative.*

Proof Recall from Example 1.6 that for any $d \in D$ and $w \in \overline{\mathbb{Q}}_{\geq 0}$, we define the unary cost function μ_w^d as follows:

$$\mu_w^d(x) \stackrel{\text{def}}{=} \begin{cases} w & \text{if } x = d, \\ 0 & \text{if } x \neq d. \end{cases}$$

Let $\mathcal{F} : D^k \to D^k$ be a non-conservative function. In that case, there are $u_1, \ldots, u_k, v_1, \ldots, v_k \in D$ such that $\mathcal{F}(u_1, \ldots, u_k) = \langle v_1, \ldots, v_k \rangle$ and there is an i such that v_i occurs more often in $\langle v_1, \ldots, v_k \rangle$ than in $\langle u_1, \ldots, u_k \rangle$. But then $\mathcal{F}$ is not a multimorphism of the unary cost function $\mu_1^{v_i}$. Hence any $\mathcal{F} \in \mathsf{Mul}(\Gamma)$ must be conservative. $\square$

Notation 5.2 For any two tuples $\mathbf{x} = \langle x_1, \ldots, x_k \rangle$ and $\mathbf{y} = \langle y_1, \ldots, y_k \rangle$ over D, we denote by $H(\mathbf{x}, \mathbf{y})$ the *Hamming distance* between $\mathbf{x}$ and $\mathbf{y}$, which is the number of positions at which the corresponding values are different.

Definition 5.3 We denote by $\Gamma_{sub,2}^{\infty}$ the set of binary submodular cost functions taking finite or infinite costs.

Theorem 5.2 *For any Boolean domain D, and any $\mathcal{F} : D^k \to D^k$, the following are equivalent*:

1. $\mathcal{F} \in \mathsf{Mul}(\Gamma_{sub,2})$.
2. $\mathcal{F} \in \mathsf{Mul}(\Gamma_{sub,2}^{\infty})$.
3. $\mathcal{F}$ is conservative and Hamming distance nonincreasing.

Proof First we consider unary cost functions. All unary cost functions on a Boolean domain are easily shown to be submodular. Hence, by Lemma 5.1, all multimorphisms of $\Gamma_{sub,2}$ and $\Gamma_{sub,2}^{\infty}$ are conservative. On the other hand, any conservative function $\mathcal{F} : D^k \to D^k$ is clearly a multimorphism of any unary cost function, since it merely permutes its arguments.

Recall from Example 1.6 that for any $w \in \overline{\mathbb{Q}}_{\geq 0}$, we define binary cost function λ_w^d as follows:

$$\lambda_w(x, y) \stackrel{\text{def}}{=} \begin{cases} c & \text{if } x = 0 \text{ and } y = 1, \\ 0 & \text{otherwise.} \end{cases}$$

Furthermore, for any $w \in \overline{\mathbb{Q}}_{\geq 0}$, we define:

$$\chi_w(x, y) \stackrel{\text{def}}{=} \begin{cases} w & \text{if } x \neq y, \\ 0 & \text{otherwise.} \end{cases}$$

Note that $\chi_w(x, y) = (\lambda_w(x, y) + \lambda_w(y, x))/2$.

By a simple case analysis, it is straightforward to check that any binary submodular cost function on a Boolean domain can be expressed by binary functions of the form λ_w, with $w > 0$, together with unary cost functions of the form μ_w^d.

We observe that when $w < \infty$, $\lambda_w(x, y) = (\chi_w(x, y) + \mu_w^0(x) + \mu_w^1(y) - w)/2$, so λ_w can be expressed by functions of the form χ_w together with unary cost functions of the form μ_w^d. Hence, since expressibility preserves multimorphisms, $\mathsf{Mul}(\Gamma_{\mathsf{sub},2}) = \mathsf{Mul}(\{\chi_w \mid w \in \mathbb{Q}_{\geq 0}, w > 0\}) \cap \mathsf{Mul}(\{\mu_w^d \mid w \in \mathbb{Q}_{\geq 0}, d \in D\})$.

Now let $\mathbf{u}, \mathbf{v} \in D^k$, and consider the multimorphism inequality, as given in Definition 1.8, for the case where $t_i = \langle \mathbf{u}[i], \mathbf{v}[i] \rangle$, for $i = 1, \ldots, k$. By Definition 1.8, for any $c > 0$, $\mathcal{F}$ is a multimorphism of χ_w if, and only if, the following holds for all choices of $\mathbf{u}$ and $\mathbf{v}$:

$$H(\mathbf{u}, \mathbf{v}) \geq H\big(\mathcal{F}(\mathbf{u}), \mathcal{F}(\mathbf{v})\big).$$

This proves that the multimorphisms of $\Gamma_{\mathsf{sub},2}$ are precisely the conservative functions that are also Hamming distance nonincreasing.

Since $\Gamma_{\mathsf{sub},2} \subseteq \Gamma_{\mathsf{sub},2}^\infty$, we know that $\mathsf{Mul}(\Gamma_{\mathsf{sub},2}^\infty) \subseteq \mathsf{Mul}(\Gamma_{\mathsf{sub},2})$. Therefore, in order to complete the proof it is enough to show that every conservative and Hamming distance nonincreasing function $\mathcal{F}$ is a multimorphism of λ_∞ and χ_∞. However, as χ_∞ is expressible by λ_∞ because $\chi_\infty(x, y) = \lambda_\infty(x, y) + \lambda_\infty(y, x)$, and expressibility preserves multimorphisms, it is enough to show that every conservative and Hamming distance nonincreasing function $\mathcal{F}$ is a multimorphism of λ_∞.

For any $\mathbf{u}, \mathbf{v} \in \{0, 1\}^k$, the Hamming distance $H(\mathbf{u}, \mathbf{v})$ is equal to the symmetric difference of the sets of positions where $\mathbf{u}$ and $\mathbf{v}$ take the value 1. Hence, for tuples $\mathbf{u}$ and $\mathbf{v}$ containing some fixed number of 1s, the minimum Hamming distance occurs precisely when one of these sets of positions is contained in the other.

Now consider again the multimorphism inequality, as given in Definition 1.8, for the case where $t_i = \langle \mathbf{u}[i], \mathbf{v}[i] \rangle$, for $i = 1, \ldots, k$. If there is any position i where $\mathbf{u}[i] = 0$ and $\mathbf{v}[i] = 1$, then $\lambda_\infty(t_i) = \infty$, so the multimorphism inequality is trivially satisfied. If there is no such position, then the set of positions where $\mathbf{v}$ takes the value 1 is contained in the set of positions where $\mathbf{u}$ takes the value 1, so $H(\mathbf{u}, \mathbf{v})$ takes its minimum possible value over all reorderings of $\mathbf{u}$ and $\mathbf{v}$. Hence if $\mathcal{F}$ is conservative, then $H(\mathbf{u}, \mathbf{v}) \leq H(\mathcal{F}(\mathbf{u}), \mathcal{F}(\mathbf{v}))$, and if $\mathcal{F}$ is Hamming distance nonincreasing, we have $H(\mathbf{u}, \mathbf{v}) = H(\mathcal{F}(\mathbf{u}), \mathcal{F}(\mathbf{v}))$. But this implies that the set of positions where

$\mathcal{F}(\mathbf{v})$ takes the value 1 is contained in the set of positions where $\mathcal{F}(\mathbf{u})$ takes the value 1. By definition of λ_∞, this implies that both sides of the multimorphism inequality are 0, so $\mathcal{F}$ is a multimorphism of λ_∞. $\square$

Remark 5.2 Together with Dave Cohen, we have observed that Mul($\Gamma_{\text{sub},2}$) are precisely those functions that are strictly subset preserving.

We now show that fractional polymorphisms of binary submodular cost functions have to be fractionally conservative.

Notation 5.3 Let $\mathcal{F} = \{\langle r_1, f_1\rangle, \ldots, \langle r_n, f_n\rangle\}$ be a k-ary fractional operation, where each r_i is a positive rational number such that $\sum_{i=1}^{n} r_i = k$ and each f_i is a distinct function from D^k to D. Then $\mathcal{F}$ is called *fractionally conservative* if for each possible choice of $x_1, \ldots, x_k$ and every $1 \leq i \leq k$,

$$\sum_{\{j \mid f_j(x_1,\ldots,x_k)=x_i\}} r_j = 1. \tag{5.1}$$

Lemma 5.2 *Let Γ be a valued constraint language including all unary cost functions. Then any fractional polymorphism $\mathcal{F}$ of Γ, $\mathcal{F} \in$ fPol(Γ), is fractionally conservative.*

Proof Recall from Example 1.6 (and the proof of Lemma 5.1) the definition of μ_w^d for any $d \in D$ and any $w \in \overline{\mathbb{Q}}_{\geq 0}$.

First notice that for every $x_1, \ldots, x_k$ and $1 \leq i \leq n$, $f_i(x_1, \ldots, x_k) \in \{x_1, \ldots, x_k\}$. (This is a weaker condition than $\mathcal{F}$ being conservative.) Otherwise, $\mathcal{F}$ cannot be a fractional polymorphism of the unary cost function $\mu_1^{f_i(x_1,\ldots,x_k)}$.

Now assume that there is an i, $1 \leq i \leq k$, such that $\sum_{\{j \mid f_j(x_1,\ldots,x_k)=x_i\}} r_j > 1$. Then clearly $\mathcal{F}$ cannot be a fractional polymorphism of the unary cost function $\mu_1^{x_i}$. Therefore, for all i, $\sum_{\{j \mid f_j(x_1,\ldots,x_k)=x_i\}} r_j \leq 1$.

Equality (5.1) now follows from the fact that for every $x_1, \ldots, x_k$,

$$\sum_{i=1}^{n} r_i = \sum_{i=1}^{k} \sum_{\{j \mid f_j(x_1,\ldots,x_k)=x_i\}} r_j = k.$$
                                                                                                                          $\square$

We now prove an extension of Theorem 5.2 to fractional polymorphisms, and hence characterise the fractional clone of binary submodular cost functions.

Notation 5.4 For any two tuples $\mathbf{x} = \langle x_1, \ldots, x_k\rangle$ and $\mathbf{y} = \langle y_1, \ldots, y_k\rangle$ over D, and a weight vector $w = \langle w_1, \ldots, w_k\rangle$, we denote by $H(\mathbf{x}, \mathbf{y}, w)$ the *weighted Hamming distance* between $\mathbf{x}$ and $\mathbf{y}$, which is the sum of all w_i such that $x_i \neq y_i$.

Theorem 5.3 *For any Boolean domain D, and any k-ary fractional operation $\mathcal{F} = \{\langle r_1, f_1\rangle, \ldots, \langle r_n, f_n\rangle\}$, where each r_i is a positive rational number such that $\sum_{i=1}^{n} r_k = k$ and each f_i is a distinct function from D^k to D, the following are equivalent:*

1. $\mathcal{F} \in \mathsf{fPol}(\Gamma_{\mathsf{sub},2})$.
2. $\mathcal{F} \in \mathsf{fPol}(\Gamma_{\mathsf{sub},2}^{\infty})$.
3. $\mathcal{F}$ is *fractionally conservative and satisfies, for any* $\mathbf{x}$ *and* $\mathbf{y}$, $H(\mathbf{x}, \mathbf{y}) \geq H(\mathcal{F}(\mathbf{x}), \mathcal{F}(\mathbf{y}), w)$, *where* $w = \langle r_1, \ldots, r_n \rangle$ (*that is,* $\mathcal{F}$ *is weighted Hamming distance nonincreasing*).

Proof Similar to the proof of Theorem 5.2. First we consider unary cost functions. All unary cost functions on a Boolean domain are easily shown to be submodular. By Lemma 5.2, all fractional polymorphisms of $\Gamma_{\mathsf{sub},2}$ and $\Gamma_{\mathsf{sub},2}^{\infty}$ are fractionally conservative. On the other hand, any fractionally conservative fractional operation $\mathcal{F}$ is clearly a fractional polymorphism of any unary cost function.

Recall from the proof of Theorem 5.2 the definition of the cost functions λ_w and χ_w for any $w \in \overline{\mathbb{Q}}_{\geq 0}$. We have seen in the proof of Theorem 5.2 that $\chi_w(x, y) = (\lambda_w(x, y) + \lambda_w(y, x))/2$, and also that any binary submodular cost function on a Boolean domain can be expressed by binary functions of the form λ_w, with $w \in \overline{\mathbb{Q}}_{\geq 0}$, together with unary cost functions of the form μ_w^d.

For $w < \infty$, $\lambda_w(x, y) = (\chi_w(x, y) + \mu_w^0(x) + \mu_w^1(y) - w)/2$; so λ_w can be expressed by functions of the form χ_w together with unary cost functions of the form μ_w^d. Hence, since expressibility preserves fractional polymorphisms, $\mathsf{fPol}(\Gamma_{\mathsf{sub},2}) = \mathsf{fPol}(\{\chi_w \mid w \in \mathbb{Q}_{\geq 0}, w > 0\}) \cap \mathsf{fPol}(\{\mu_w^d \mid w \in \mathbb{Q}_{\geq 0}, d \in D\})$.

Now let $\mathbf{u}, \mathbf{v} \in D^k$, and consider the fractional polymorphism Inequality (1.3), as given in Definition 1.10, for the case where $t_i = \langle \mathbf{u}[i], \mathbf{v}[i] \rangle$, for $i = 1, \ldots, k$. Let $p = H(\mathbf{u}, \mathbf{v})$, and let $q = H(\mathbf{u}, \mathbf{v}, w)$, where $w = \langle r_1, \ldots, r_n \rangle$. By Definition 1.10, $\mathcal{F}$ is a fractional polymorphism of χ_w if, and only if, $p \geq q$.

This proves that the fractional polymorphisms of $\Gamma_{\mathsf{sub},2}$ are precisely the fractionally conservative fractional operations that are weighted Hamming distance nonincreasing.

Since $\Gamma_{\mathsf{sub},2} \subseteq \Gamma_{\mathsf{sub},2}^{\infty}$, we know that $\mathsf{fPol}(\Gamma_{\mathsf{sub},2}^{\infty}) \subseteq \mathsf{fPol}(\Gamma_{\mathsf{sub},2})$. Using an argument similar to the one in the proof of Theorem 5.2, in order to complete the proof it is enough to show that every fractional operation that is fractionally conservative and weighted Hamming distance nonincreasing is a fractional polymorphism of λ_{∞}.

Let $\mathbf{u}, \mathbf{v} \in D^k$. We denote $t_i = \langle \mathbf{u}[i], \mathbf{v}[i] \rangle$, for $i = 1, \ldots, k$, and we denote $t_i' = \langle \mathcal{F}(\mathbf{u})[i], \mathcal{F}(\mathbf{v})[i] \rangle$, for $i = 1, \ldots, n$. If there is any position i where $\mathbf{u}[i] = 0$ and $\mathbf{v}[i] = 1$, then $\lambda_{\infty}(t_i) = \infty$, and the fractional polymorphism Inequality (1.3), as given in Definition 1.10, is trivially satisfied. Hence we can assume that there is no such position. We denote by x_{00} the number of $\langle 0, 0 \rangle$ tuples among t_i, $1 \leq i \leq k$, and similarly for x_{01}, x_{10}, and x_{11}. (Note that by our assumption, $x_{01} = 0$.) We denote by w_{00} the sum of weights w_i where $t_i' = \langle 0, 0 \rangle$, and similarly for w_{01}, w_{10}, and w_{11}.

Since $\mathcal{F}$ is fractionally conservative, we get the following:

$$x_{00} = w_{00} + w_{01}, \tag{5.2}$$

$$x_{00} + x_{10} = w_{00} + w_{10}. \tag{5.3}$$

Moreover, since $\mathcal{F}$ is weighted Hamming distance nonincreasing, we get the following:

$$x_{10} \geq w_{01} + w_{10}. \tag{5.4}$$

Equation (5.3) and Inequality (5.4) give

$$w_{00} - x_{00} \geq w_{01}. \tag{5.5}$$

Equation (5.2) and Inequality (5.5) give

$$0 \geq w_{01}. \tag{5.6}$$

But this means that there are no $\langle 0, 1 \rangle$ tuples among t_i', $1 \leq i \leq k$. By definition of λ_∞, this implies that both sides of the fractional polymorphism Inequality (1.3), as given in Definition 1.10, are 0, so $\mathcal{F}$ is a fractional polymorphism of λ_∞. $\square$

5.5 Non-expressibility of Γ_{sub} over $\Gamma_{\text{sub},2}$

Theorem 5.2 characterises the multimorphisms of $\Gamma_{\text{sub},2}$, and hence enables us to systematically search (for example, using MATHEMATICA) for multimorphisms of $\Gamma_{\text{sub},2}$ that are not multimorphisms of Γ_{sub}. In this way, we have identified the function $\mathcal{F}_{\text{sep}} : \{0, 1\}^5 \to \{0, 1\}^5$ defined in Fig. 5.1. We will show in this section that this function has the remarkable property that it can be used to characterise all the submodular functions of arity 4 that are expressible by binary submodular functions on a Boolean domain. Using this result, we show that some submodular functions are *not* expressible in this way, because they do not have $\mathcal{F}_{\text{sep}}$ as a multimorphism.

Proposition 5.1 $\mathcal{F}_{\text{sep}}$ *is conservative and Hamming distance nonincreasing.*

Proof Straightforward exhaustive verification. $\square$

Theorem 5.4 *For any function $f \in \Gamma_{\text{sub},4}$ the following are equivalent:*

1. $f \in \langle \Gamma_{\text{sub},2} \rangle$.
2. $\mathcal{F}_{\text{sep}} \in \text{Mul}(\{f\})$.
3. $f \in \text{Cone}(\Gamma_{\text{fans},4})$.

Proof First, we show (1) $\Rightarrow$ (2). Proposition 5.1 and Theorem 5.2 imply that $\mathcal{F}_{\text{sep}}$ is a multimorphism of any binary submodular function on a Boolean domain. Hence having $\mathcal{F}_{\text{sep}}$ as a multimorphism is a necessary condition for any submodular cost function on a Boolean domain to be expressible by binary submodular cost functions.

Next, we show (2) $\Rightarrow$ (3). Consider the complete set of inequalities on the values of a 4-ary cost function resulting from having the multimorphism $\mathcal{F}_{\text{sep}}$, as specified in Definition 1.8. A routine calculation in MATHEMATICA shows that, out of 16^5 such inequalities, there are 4,635 that are distinct. After removing from these all those that are equal to the sum of two others, we obtain a system of just 30 inequalities which must be satisfied by any 4-ary submodular cost function that has

$$
\begin{array}{ll}
\mathbf{x} &
\begin{cases}
00000000000000001111111111111111 \\
00000000111111110000000011111111 \\
00001111000011110000111100001111 \\
00110011001100110011001100110011 \\
01010101010101010101010101010101
\end{cases} \\[2pt]
\hline \\[-6pt]
\mathcal{F}_{\mathrm{sep}}(\mathbf{x}) &
\begin{cases}
00000000000000000001000100010001 \\
00000000000001010000000000000111 \\
00000011000100110000011111111111 \\
00010101111111111111111111111111 \\
01111111011111110111111101111111
\end{cases}
\end{array}
$$

Fig. 5.1 Definition of $\mathcal{F}_{\mathrm{sep}}$

the multimorphism $\mathcal{F}_{\mathrm{sep}}$. Using the double description method [219],[2] we obtain from these 30 inequalities an equivalent set of 31 extreme rays that generate the same polyhedral cone of cost functions. These extreme rays all correspond to fans or sums of fans.

Finally, we show (3) $\Rightarrow$ (1). By Theorem 5.1, all fans are expressible over $\varGamma_{\mathrm{sub},2}$. It follows that any cost function in this cone of functions is also expressible over $\varGamma_{\mathrm{sub},2}$. $\qquad\square$

Next we show that there are indeed 4-ary submodular cost functions that do not have $\mathcal{F}_{\mathrm{sep}}$ as a multimorphism and therefore are not expressible by binary submodular cost functions.

Definition 5.4 For any Boolean tuple t of arity 4 containing exactly two 1s and two 0s, we define the 4-ary cost function θ_t as follows:

$$
\theta_t(x_1, x_2, x_3, x_4) \stackrel{\mathrm{def}}{=}
\begin{cases}
-1 & \text{if } (x_1, x_2, x_3, x_4) = (1, 1, 1, 1) \text{ or } (0, 0, 0, 0), \\
1 & \text{if } (x_1, x_2, x_3, x_4) = t, \\
0 & \text{otherwise.}
\end{cases}
$$

Cost functions of the form θ_t have been introduced in [236], where they are called *quasi-indecomposable* functions.

Definition 5.5 We denote by $\varGamma_{\mathrm{qin}}$ the set of all (six) quasi-indecomposable cost functions of arity 4.

It is straightforward to check that all cost functions from $\varGamma_{\mathrm{qin}}$ are submodular, but the next result shows that they are *not* expressible by binary submodular functions.

Proposition 5.2 *For all* $\theta \in \varGamma_{\mathrm{qin}}$, $\mathcal{F}_{\mathrm{sep}} \notin \mathsf{Mul}(\{\theta\})$.

---

[2]As implemented by the program SKELETON available from http://www.uic.nnov.ru/~zny/skeleton/.

$$\left.\begin{matrix} 1\,0\,1\,0 \\ 1\,0\,0\,1 \\ 0\,1\,0\,1 \\ 0\,1\,1\,0 \\ 0\,0\,1\,1 \end{matrix}\,\,\xrightarrow{\theta_{(1,1,0,0)}}\,\, \begin{matrix} 0 \\ 0 \\ 0 \\ 0 \\ 0 \end{matrix}\right\}\,\,\sum = 0$$

$$\mathcal{F}_{\text{sep}}\left.\begin{matrix} 0\,0\,1\,0 \\ 0\,0\,0\,1 \\ 1\,1\,0\,0 \\ 1\,0\,1\,1 \\ 0\,1\,1\,1 \end{matrix}\,\,\xrightarrow{\theta_{(1,1,0,0)}}\,\, \begin{matrix} 0 \\ 0 \\ 1 \\ 0 \\ 0 \end{matrix}\right\}\,\,\sum = 1$$

Fig. 5.2 $\mathcal{F}_{\text{sep}} \notin \text{Mul}(\{\theta_{(1,1,0,0)}\})$

Proof The tableau in Fig. 5.2 shows that $\mathcal{F}_{\text{sep}} \notin \text{Mul}(\{\theta_{(1,1,0,0)}\})$. Permuting the columns appropriately establishes the result for all other $\theta \in \Gamma_{\text{qin}}$. □

Corollary 5.1 *For all $\theta \in \Gamma_{\text{qin}}$, $\theta \notin \langle \Gamma_{\text{sub},2} \rangle$.*

Proof By Theorem 5.4 and Proposition 5.2. □

Are there any other 4-ary submodular cost functions that are not expressible over $\Gamma_{\text{sub},2}$? Promislow and Young characterised the extreme rays of the cone of all 4-ary submodular cost functions and established that $\Gamma_{\text{sub},4} = \text{Cone}(\Gamma_{\text{fans},4} \cup \Gamma_{\text{qin}})$—see Theorem 5.2 of [236]. Hence the results in this section characterise the expressibility of all 4-ary submodular functions.

Promislow and Young conjectured that for $k \neq 4$, all extreme rays of $\Gamma_{\text{sub},k}$ are fans [236]; that is, they conjectured that for all $k \neq 4$, $\Gamma_{\text{sub},k} = \text{Cone}(\Gamma_{\text{fans},k})$. However, if this conjecture were true it would imply that all submodular functions of arity 5 and above were expressible by binary submodular functions, by Theorem 5.1. This is clearly not the case, because inexpressible cost functions such as those identified in Corollary 5.1 can be extended to larger arities (for instance, by adding dummy arguments) and remain inexpressible. Hence our results refute this conjecture for all $k \geq 5$. However, we suggest that this conjecture can be refined to a similar statement concerning just those submodular functions that are expressible by binary submodular functions, as follows:

Conjecture 5.1 For all k, $\Gamma_{\text{sub},k} \cap \langle \Gamma_{\text{sub},2} \rangle = \text{Cone}(\Gamma_{\text{fans},k})$.

This conjecture was previously known to be true for $k \leq 3$ [236]; Theorem 5.1 shows that $\text{Cone}(\Gamma_{\text{fans},k}) \subseteq \Gamma_{\text{sub},k} \cap \langle \Gamma_{\text{sub},2} \rangle$ for all k, and Theorem 5.4 confirms that equality holds for $k = 4$.

Remark 5.3 We have seen in Chap. 3 that in the case of max-closed cost functions there is a difference between finite-valued and general-valued cost functions.

Adding infinite costs makes the hierarchy collapse. By contrast, in the case of submodular cost functions, there is no difference between finite-valued and general-valued cost functions. There are finite-valued submodular cost functions that are not expressible, namely cost functions from Γ_{qin}, and even adding infinite costs to the hidden variables would not help to make them expressible: by Theorem 1.3, expressive power is characterised by fractional polymorphisms; by Theorem 5.3, finite-valued and general-valued submodular cost functions have the same fractional polymorphisms.

5.6 The Complexity of Recognising Expressible Functions

Finally, we show that we can test efficiently whether a submodular polynomial of arity 4 is expressible by binary submodular polynomials.

Definition 5.6 Let $p(x_1, x_2, x_3, x_4)$ be the polynomial representation of a 4-ary submodular cost function f. We denote by a_I the coefficient of the term $\prod_{i \in I} x_i$. We say that f satisfies condition **Sep** if for each $\{i, j\}, \{k, \ell\} \subset \{1, 2, 3, 4\}$, with i, j, k, ℓ distinct, we have $a_{\{i,j\}} + a_{\{k,\ell\}} + a_{\{i,j,k\}} + a_{\{i,j,\ell\}} \leq 0$.

Theorem 5.5 For any $f \in \Gamma_{\text{sub},4}$, the following are equivalent:

1. $f \in \langle \Gamma_{\text{sub},2} \rangle$.
2. f satisfies condition **Sep**.

Proof As in the proof of Theorem 5.4, we construct a set of 30 inequalities corresponding to the multimorphism $\mathcal{F}_{\text{sep}}$. Each of these inequalities on the values of a cost function can be translated into inequalities on the coefficients of the corresponding polynomial representation by a straightforward linear transformation. This calculation shows that 24 of the resulting inequalities impose the condition of submodularity, and the remaining six impose condition **Sep**. Hence a submodular cost function of arity 4 has the multimorphism $\mathcal{F}_{\text{sep}}$ if, and only if, its polynomial representation satisfies condition **Sep**. The result then follows from Theorem 5.4. $\square$

Using Theorem 5.5, we can test whether optimisation problems given as a sum of submodular functions of arity 4 can be reduced to the (s, t)-MIN-CUT problem via the expressibility reduction. (These problems arise in computer vision and in valued constraint satisfaction problems and will be mentioned in Sect. 5.7.)

Furthermore, by Theorem 5.1, the number of extra variables needed in this reduction is rather small compared to the theoretical upper bound given in Proposition 2.2 [65].

It is known that the problem of recognising whether an arbitrary degree-4 polynomial is submodular is co-NP-complete [83, 124]. One might hope that the more restricted class of submodular polynomials expressible by binary submodular polynomials would be recognisable in polynomial time. At the moment, the complexity of the recognition problem for submodular polynomials of degree 4 that are expressible by binary submodular polynomials is open.

Remark 5.4 Multimorphism $\mathcal{F}_{sep}$ could be used to test whether a given polynomial p of degree 4 is expressible by binary submodular polynomials (and is therefore submodular). However, the only known way of testing whether $\mathcal{F}_{sep}$ is a multimorphism of a given polynomial p (in n variables) is via testing all possible tableaux; this would take exponential time in n.

Remark 5.5 Martin Cooper noticed that the easier problem of testing whether a polynomial (in n variables) of degree 4 can be written as the sum of 4-ary submodular polynomials (that is, without any extra variables) can be solved in polynomial time just by solving a system of linear equations.

Consequently, assuming $P \neq$ co-NP, there are submodular polynomials of degree 4 that cannot be written as the sum of 4-ary submodular polynomials, and hence need extra variables to be expressible. However, this is not surprising: we have seen an example of such a polynomial in Example 4.3.

5.7 Summary

In this section, we have extended our study of the expressive power of binary submodular cost functions. We showed a new class of submodular cost functions, the so-called *fans*, which are expressible by binary submodular cost functions. We also showed that there are submodular cost functions that are not expressible by binary submodular cost functions, and hence are not minimisable by a reduction to (s, t)-MIN-CUT via the expressibility reduction. Moreover, we characterised precisely which submodular cost functions of arity 4 can be expressed by binary submodular cost functions. We also presented results on the recognition problem, and some more results on the algebraic properties of submodular cost functions. We finish this chapter with applications of our results, and remarks on related work.

5.7.1 Applications to Artificial Intelligence

As mentioned in Chap. 4, any Boolean cost function of arity k can be represented uniquely as a Boolean polynomial. Moreover, if Γ is a set of cost functions on a Boolean domain, with arity at most k, then any instance of VCSP(Γ) with n variables can be uniquely represented as a polynomial p in n Boolean variables, of degree at most k. Conversely, any such polynomial represents an n-ary cost function that can be expressed over a set of cost functions on a Boolean domain, with arity at most k. Note also that over a Boolean domain we have that $x^2 = x$, so p has at most 2^n terms: these correspond to subsets of variables.

Example 5.7 Any unary cost function ϕ on a Boolean domain can be expressed as the polynomial $p(x_1) = \phi(0) + (\phi(1) - \phi(0))x_1$. Similarly, any binary cost function ϕ can be expressed as

$$p(x_1, x_2) = \phi(0, 0)$$
$$+ \left(\phi(1, 0) - \phi(0, 0)\right)x_1$$
$$+ \left(\phi(0, 1) - \phi(0, 0)\right)x_2$$
$$+ \left(\phi(1, 1) - \phi(0, 1) - \phi(1, 0) + \phi(0, 0)\right)x_1 x_2.$$

Consider the valued constraint language Γ from Example 1.7. It is easy to check that all of the cost functions in Γ are submodular: unary cost functions are submodular by definition; each binary $\phi \in \Gamma$ satisfies $\phi(0, 0) + \phi(1, 1) \leq \phi(0, 1) + \phi(1, 0)$. Hence the instance $\mathcal{P}$ from Example 1.7 is an instance of $\mathsf{VCSP}(\Gamma_{\mathsf{sub},2})$. The corresponding polynomial is

$$p(x_1, \ldots, x_5) = 3 + 0x_1 - x_2 - x_1 x_2$$
$$+ 0 + 2x_1 + 4x_4 - x_1 x_4$$
$$+ 0 + 0x_2 + x_3 - x_2 x_3$$
$$+ 9 - x_3 - 2x_4 - 5x_3 x_4$$
$$+ 3 + x_3 + 2x_5 - 2x_3 x_5$$
$$+ 4 - 2x_4 - x_5 + 0x_4 x_5$$
$$+ 0 + 5x_2$$
$$+ 4 - 2x_5,$$

which can be simplified to give

$$p(x_1, \ldots, x_5) = 23 + 2x_1 + 4x_2 + x_3 - x_5$$
$$- x_1 x_2 - x_1 x_4 - x_2 x_3 - 5x_3 x_4 - 2x_3 x_5.$$

The fact that $\mathcal{P}$ is an instance of $\mathsf{VCSP}(\Gamma_{\mathsf{sub},2})$ can be easily seen from the polynomial representation: the polynomial p has all quadratic coefficients nonpositive, and hence is submodular.

We can rewrite p as in the proof of Theorem 4.1 as follows:

$$p(x_1, \ldots, x_5) = 23 + 2x_1 + 4x_2 + x_3 - x_5$$
$$+ (1 - x_1)x_2 - x_2 + (1 - x_1)x_4 - x_4 + (1 - x_2)x_3 - x_3$$
$$+ 5(1 - x_3)x_4 - 5x_4 + 2(1 - x_3)x_5 - 2x_5$$
$$= 23 + 2x_1 + 3x_2 - 6x_4 - 3x_5$$
$$+ (1 - x_1)x_2 + (1 - x_1)x_4$$
$$+ (1 - x_2)x_3 + 5(1 - x_3)x_4 + 2(1 - x_3)x_5$$
$$= 14 + 2x_1 + 3x_2 + 6(1 - x_4) + 3(1 - x_5)$$
$$+ (1 - x_1)x_2 + (1 - x_1)x_4$$
$$+ (1 - x_2)x_3 + 5(1 - x_3)x_4 + 2(1 - x_3)x_5.$$

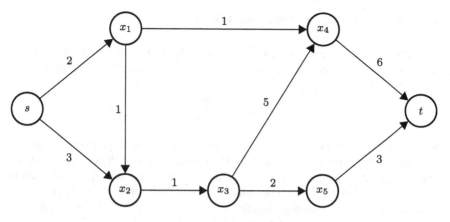

Fig. 5.3 Graph G corresponding to polynomial p (Example 5.7)

We can now build a graph G with five vertices corresponding to variables x_1 through x_5 and two extra vertices s and t and add edges accordingly (see Fig. 5.3).

For every assignment v of values 0 and 1 to variables x_1, x_2, x_3, x_4, x_5, $p(v(x_1), \ldots, v(x_5))$ is equal to the size of the (s, t)-cut in G given by v plus 14 (for the constant term in the posiform representation of p). The minimum cut in G, with value 2, is the set $\{s, x_1, x_2\}$. Therefore, the assignment $(x_1 \mapsto 0, x_2 \mapsto 0, x_3 \mapsto 1, x_4 \mapsto 1, x_5 \mapsto 1)$ minimises the polynomial p with total value 16.

Using the same method as in Example 5.7, we obtain:

Corollary 5.2 (of Theorem 5.1) *For any fixed $k \geq 4$, any* VCSP$(\Gamma_{\mathsf{fans},k})$ *instance with n variables and p_i constraints of arity i, $3 \leq i \leq k$, is solvable in $O((n + p)^3)$ time, where $p = \sum_{i=3}^{k} p_i(1 + \lfloor i/2 \rfloor)$.*

As shown above, VCSP$(\Gamma_{\mathsf{fans},4})$ is the *maximal* class in VCSP$(\Gamma_{\mathsf{sub},4})$ that can be solved by reduction to (s, t)-MIN-CUT in this way.

Cohen et al. have shown [65] that if a cost function ϕ of arity k is expressible by some set of cost functions over Γ, then ϕ is expressible by Γ using at most 2^{2^k} extra variables (Proposition 2.2). Theorem 5.1 shows that only $O(k)$ extra variables are needed to express any cost function from $\Gamma_{\mathsf{fans},k}$ by $\Gamma_{\mathsf{sub},2}$. Therefore, instances of VCSP(Γ_{fans}) need fewer extra variables than the theoretical upper bound given by Proposition 2.2. In particular, an instance of VCSP$(\Gamma_{\mathsf{sub},4})$ is either reducible to (s, t)-MIN-CUT with only linearly many extra variables,[3] or is not reducible at all.

---

[3]Optimal (in the number of extra variables) gadgets for all cost functions from $\Gamma_{\mathsf{fans},4}$ have been identified in [278].

5.7.2 Applications to Computer Vision

In computer vision, many problems can be naturally formulated in terms of *energy minimisation* where the energy function, over a set of variables $\{x_v\}_{v \in V}$, has the following form:

$$E(\mathbf{x}) = c_0 + \sum_{v \in V} c_v(x_v) + \sum_{\langle u,v \rangle \in V \times V} c_{uv}(x_u, x_v) + \cdots .$$

Set V usually corresponds to pixels: x_v denotes the label of pixel $v \in V$, which must belong to some finite domain D. The constant term of the energy is c_0; the unary terms $c_v(\cdot)$ encode data penalty functions; the pairwise and higher-order terms $c_{uv}(\cdot, \cdot)$ and so on are interaction potentials. Functions of arity 3 and above are known as higher-order energy functions, or higher-order cliques. This energy is often derived in the context of MARKOV RANDOM FIELDS (also CONDITIONAL RANDOM FIELD) [18, 128]: a minimum of E corresponds to a *maximum a posteriori* (MAP) labelling $\mathbf{x}$ [203, 267].

As discussed above, there is a direct translation between VCSP instances and pseudo-Boolean polynomials. Hence it is clear that the above-mentioned framework of energy optimisation is equivalent to the VCSP. (See [268] for a survey on the connection between computer vision and constraint satisfaction problems, although with a strong emphasis on a linear programming approach.) Therefore, for energy minimisation over Boolean variables we get the following:

Corollary 5.3 (of Theorem 5.1) *Energy minimisation, where each energy function belongs to Γ_{fans}, is solvable in $O((n + p)^3)$ time, where n is the number of variables (pixels), p_i is the number of energy functions of arity i and $p = \sum_{i=3}^{k} p_i(1 + \lfloor i/2 \rfloor)$.*

Higher-order energy functions have the ability to encode high-level structural dependencies between pixels, which have been shown to be extremely powerful for image labelling problems. They have long been used to model image textures [200, 228, 249], image denoising and restoration [249], and texture segmentation [181]. Their use, however, is severely hampered in practice by the intractable complexity of representing and minimising such functions [250]. Our results enlarge the class of higher-order energy functions that can be (exactly) minimised efficiently using graphs cuts. Moreover, despite the theoretical double-exponential upper bound on the number of extra variables (Proposition 2.2), the proof of Theorem 5.1 shows that any function from Γ_{fans} needs only linearly many (in the arity of the function) extra variables. Hence functions from Γ_{fans} could be used, for instance, in image processing for efficient recognition of images or in Bayesian estimation.

5.7.3 Related Work

We describe an alternative approach to the question of expressibility of submodular cost functions. Cohen et al. have shown that a VCSP instance with binary submodular constraints over a totally-ordered domain can be minimised via (s, t)-MIN-CUT in cubic time [58]. Furthermore, Cohen et al. have shown that a VCSP instance with so-called *2-monotone* constraints over a lattice-ordered domain, which form a subclass of submodular constraints, can be solved via (s, t)-MIN-CUT in cubic time. Given a submodular Boolean VCSP instance $\mathcal{P}$ with constraints of arity at most r, one could try to use the *dual representation* [98, 202], which transforms $\mathcal{P}$ into a binary VCSP instance with an exponential blowup (in r) of the domain size. One could hope to combine these two results and obtain an algorithm for submodular Boolean VCSPs using the (s, t)-MIN-CUT problem. An obvious way to complete the above-mentioned construction would be to prove that every binary submodular cost function over a lattice-ordered set can be expressed by 2-monotone cost functions. However, this is not possible as Corollary 5.1 gives an example of a binary submodular cost function defined over a lattice-ordered set (in fact, in this case the lattice is distributive) that is not expressible by binary submodular cost functions (and hence not by 2-monotone cost functions).

Given the limitations of reducing the SFM_b problem to the (s, t)-MIN-CUT problem, as explained in this section, other approaches have been suggested. Kolmogorov formulated the SFM_b problem using submodular flows [186]. Our results have been used in follow-up work [52, 178].

For more on quadratisation of pseudo-Boolean functions, not necessarily preserving submodularity, see [33].

The question of which submodular functions can be "realised" (expressed without introducing extra variables) as graph cut functions was considered in [123].

5.7.4 Open Problems

Functions which can be made polar by switching a subset of variables are called *unimodular*. The class of unimodular functions was studied in [148], and shown to be polynomial-time recognisable in [83]. Note that unimodular functions are not in general submodular. It would be nice to know the precise relationship between submodular cost functions that are expressible by binary submodular cost functions and unimodular cost functions that are submodular. Moreover, what is the largest subclass of submodular cost functions that is polynomial-time recognisable?

Is there an infinite hierarchy of submodular cost functions of increasing expressive power (like finite-valued max-closed cost functions; cf. Chap. 3) or are all submodular cost functions expressible by a subset of submodular cost functions of a fixed arity?

Is there any characterisation of all separating multimorphisms of expressible submodular cost functions of arity 4 (we have shown just one such multimorphism, namely $\mathcal{F}_{\text{sep}}$) or, more generally, of submodular cost functions of arbitrary arities?

What is the precise boundary between expressible and non-expressible submodular cost functions (cf. Conjecture 5.1)?

What is the complexity of the recognition problem for submodular polynomials that are expressible by binary submodular polynomials?

Is there any connection between fans and the general polynomial-time algorithm for the SFM problem by Iwata and Orlin [160]?

Our work deals with Boolean submodular cost functions only. Is there any kind of norm which characterises the algebraic properties of non-Boolean submodular cost functions as Hamming distance does in the case of Boolean submodular cost functions?

Part III
Tractability

Chapter 6
Tractable Languages

> *I learned very early the difference between knowing the name of something and knowing something.*
> Richard Feynman

6.1 Introduction

This chapter, based on [168], presents a list of known tractable languages. Quite strikingly, all of them can be characterised by simple fractional polymorphisms.

6.2 Tractable Languages

We start with a simple tractable language characterised by a unary multimorphism.

Example 6.1 (c-validity) Given a fixed $c \in D$, we say that language Γ is c-valid if for all $\phi \in \Gamma$, $\phi(c, \ldots, c) \leq \phi(\mathbf{x})$ holds for all $\mathbf{x} \in D^m$, where m is the arity of ϕ. It is clear that c-valued languages are tractable, as assigning the value c to all variables yields an optimal solution.

It is easy to see that Γ is c-valid if, and only if, Γ admits $\langle g_c \rangle$ as a multimorphism, where $g_c : D \to D$ is defined by $g_c(x) = c$ for all $x \in D$ (that is, g_c is the constant function returning the value c).

We now move on to tractable languages characterised by binary multimorphisms.

Example 6.2 (Min) Let $\Gamma_{\min}$ be a language, over some totally-ordered domain D, that is improved by the pair $\langle \text{Min}, \text{Min} \rangle$, where $\text{Min} : D^2 \to D$ is the binary operation returning the smaller of its two arguments. For crisp languages, the tractability of $\Gamma_{\min}$ was shown in [167]. Using this result, it was shown in [67] that after establishing arc consistency [248] (which might restrict the domains of individual variables), assigning the smallest remaining domain value to all variables will yield an optimal solution. Thus $\Gamma_{\min}$ is tractable.

S. Živný, *The Complexity of Valued Constraint Satisfaction Problems*,
Cognitive Technologies, DOI 10.1007/978-3-642-33974-5_6,
© Springer-Verlag Berlin Heidelberg 2012

Example 6.3 (Max) Let Γ_{max} be a language, over some totally-ordered domain D, that is improved by the pair $\langle Max, Max \rangle$, where $Max : D^2 \to D$ is the binary operation returning the larger of its two arguments. The tractability of Γ_{max} has been shown in [67], using a similar argument to the one in Example 6.2.

Our next example, already discussed in Sect. 1.6, covers the well-studied concept of submodularity.

Example 6.4 (Submodularity) Let Γ_{sub} be a language, over some totally ordered domain D, that admits $\langle Min, Max \rangle$ as a multimorphism (cf. Definition 1.12), where Min and Max are defined as in Examples 6.2 and 6.3. Using a polynomial-time strongly combinatorial algorithm for minimising submodular functions, such as the algorithms described in Sect. 1.6, it was shown in [67] that Γ_{sub} is tractable.

Some special cases of Γ_{sub} admit more efficient algorithms. For example, in the case when $D = \{0, 1\}$, the cost functions defined in Example 1.6 are all submodular, so the language Γ_{cut} defined in Example 1.6 is strictly contained in Γ_{sub} for D. (Indeed, it is well known that the cut function for any graph is submodular.) Since cut functions can be minimised more efficiently than general submodular functions (cf. Example 1.8), classes of submodular functions from Γ_{sub} that are expressible over Γ_{cut} have been studied [274, 280]. However, it has been shown in [274, 275] that not all functions from Γ_{sub} are expressible over Γ_{cut}, so this approach cannot be used to obtain a more efficient algorithm for the whole of VCSP(Γ_{sub}).

Other approaches for solving instances from VCSP(Γ_{sub}) include linear programming [73, 75] and submodular flows [186].

All the results mentioned above deal with submodularity on a totally-ordered domain. However, submodularity has been defined over arbitrary lattices (cf. Definition 1.12). In fact, the minimisation algorithms for submodular functions (even for functions given by an oracle), mentioned in Sect. 1.6, work for submodular functions on any distributive lattice [159, 257]. A pseudo-polynomial-time algorithm for minimising submodular functions on diamonds is known [192], and several constructions on lattices preserving tractability have been identified [191].

A recent result of Thapper and Živný shows that a simple linear programme solves VCSPs with submodular cost functions on *arbitrary* lattices [264]. (More on this in Chap. 8.)

Example 6.5 (Bisubmodularity) For a given finite set V, bisubmodular functions are functions defined on pairs of disjoint subsets of V with a requirement similar to that of submodularity in Definition 1.11. In particular, given a finite nonempty set V, a function $f : 3^V \to \mathbb{Q}_{\geq 0}$ defined on pairs of disjoint subsets[1] of V is called *bisubmodular* [122] if for all pairs $\langle S_1, S_2 \rangle$ and $\langle T_1, T_2 \rangle$ of disjoint subsets of U,

$$f\big(\langle S_1, S_2 \rangle \sqcap \langle T_1, T_2 \rangle\big) + f\big(\langle S_1, S_2 \rangle \sqcup \langle T_1, T_2 \rangle\big) \leq f\big(\langle S_1, S_2 \rangle\big) + f\big(\langle T_1, T_2 \rangle\big),$$

---

[1] Think of 3^V as $\{0, 1, 2\}$-vectors of length $|V|$.

where

$$\langle S_1, S_2 \rangle \sqcap \langle T_1, T_2 \rangle = \langle S_1 \cap T_1, S_2 \cap T_2 \rangle,$$

$$\langle S_1, S_2 \rangle \sqcup \langle T_1, T_2 \rangle = \big\langle (S_1 \cup T_1) \setminus (S_2 \cup T_2), (S_2 \cup T_2) \setminus (S_1 \cup T_1) \big\rangle.$$

To give some examples, rank functions of delta-matroids are bisubmodular [35, 51]. Bisubmodularity also arises in bicooperative games [19]. The minimisation problem for bisubmodular functions can be solved using the ellipsoid method as shown in [237].

The first combinatorial algorithm for minimising an integer-valued bisubmodular function (BSFM) is due to Fujishige and Iwata [122]. The time complexity of the fastest known general strongly polynomial algorithm for BSFM is $O(n^7 EO \log n)$, where $n = |U|$ and EO is the time for function evaluation [217].

A property equivalent to bisubmodularity can be defined on cost functions on the set $D = \{0, 1, 2\}$. We define two binary operations Min_0 and Max_0 as follows:

$$\mathrm{Min}_0(x, y) \overset{\text{def}}{=} \begin{cases} 0 & \text{if } 0 \neq x \neq y \neq 0, \\ \mathrm{Min}(x, y) & \text{otherwise,} \end{cases}$$

and

$$\mathrm{Max}_0(x, y) \overset{\text{def}}{=} \begin{cases} 0 & \text{if } 0 \neq x \neq y \neq 0, \\ \mathrm{Max}(x, y) & \text{otherwise.} \end{cases}$$

We denote by Γ_{bis} the set of finite-valued cost functions (that is, with range $\mathbb{Q}_{\geq 0}$) defined on D that admit $\langle \mathrm{Min}_0, \mathrm{Max}_0 \rangle$ as a multimorphism. By the results mentioned above [122, 217], Γ_{bis} is tractable. However, this result does not imply the tractability of general-valued languages defined on $D = \{0, 1, 2\}$ admitting $\langle \mathrm{Min}_0, \mathrm{Max}_0 \rangle$ as a multimorphism. The tractability of such languages is discussed in Chap. 8.

Example 6.6 (Skew bisubmodularity) We extend the notion of bisubmodularity from Example 6.5 to skew bisubmodularity [154]. Let $D = \{0, 1, 2\}$. We define

$$\mathrm{Max}_1(x, y) \overset{\text{def}}{=} \begin{cases} 1 & \text{if } 0 \neq x \neq y \neq 0, \\ \mathrm{Max}(x, y) & \text{otherwise.} \end{cases}$$

A language Γ defined on D is called α-bisubmodular, for some real $0 < \alpha \leq 1$, if Γ admits $\{\langle 1, \mathrm{Min}_0 \rangle, \langle \alpha, \mathrm{Max}_0 \rangle, \langle 1 - \alpha, \mathrm{Max}_1 \rangle\}$ as a fractional polymorphism, where Min_0 and Max_0 are defined as in Example 6.5. Note that 1-bisubmodular languages are (ordinary) bisubmodular languages from Example 6.5. The concept of α-bisubmodular languages has been identified in [154] and shown tractable using results from [264] (more details in Chap. 8).

Example 6.7 (k-Submodularity) Let Γ be a language over a domain D, where $|D| > 3$, and admitting the multimorphism $\langle \mathrm{Min}_0, \mathrm{Max}_0 \rangle$ from Example 6.5. Such languages are known as *k-submodular* [153].[2] The tractability of (general-valued) *k*-submodular languages has recently been shown in [264] and is discussed in more detail in Chap. 8.

Example 6.8 ((Symmetric) tournament pair)   A binary operation $f : D^2 \to D$ is called a *tournament* operation if (i) f is commutative, that is, $f(x, y) = f(y, x)$ for all $x, y \in D$; and (ii) f is conservative, that is, $f(x, y) \in \{x, y\}$ for all $x, y \in D$. The *dual* of a tournament operation is the unique tournament operation g satisfying $x \neq y \Rightarrow g(x, y) \neq f(x, y)$.

A *tournament pair* is a pair $\langle f, g \rangle$, where both f and g are tournament operations. A tournament pair $\langle f, g \rangle$ is called *symmetric* if g is the dual of f.

Let Γ be an arbitrary language that admits a symmetric tournament pair multimorphism. It has been shown in [66], by a reduction to submodular languages from Example 6.4, that any such Γ is tractable.

Now let Γ be an arbitrary language that admits any tournament pair multimorphism. It has been shown in [66], by a reduction to the symmetric tournament pair case, that any such Γ is also tractable.

Our next example presents recent results on submodular languages defined over domains with a certain structure.

Example 6.9 (Tree-submodularity)   Assume that the domain values from D can be arranged into a binary tree T, that is, a tree where each node has at most two children. Given $a, b \in T$, let P_{ab} denote the unique path in T between a and b of length (= number of edges) $d(a, b)$, and let $P_{ab}[i]$ denote the ith vertex on P_{ab}, where $0 \leq i \leq d(a, b)$ and $P_{ab}[0] = a$. Let $\langle g_\sqcap, g_\sqcup \rangle$ be two binary operations satisfying $\{g_\sqcap(a, b), g_\sqcup(a, b)\} = \{P_{ab}[\lfloor d/2 \rfloor], P_{ab}[\lceil d/2 \rceil]\}$.

A language admitting $\langle g_\sqcup, g_\sqcap \rangle$ as a multimorphism has been called *strongly tree-submodular*. The tractability of strongly tree-submodular languages was shown in [185].

For $a, b \in T$, let $g_\wedge(a, b)$ be defined to be the highest common ancestor of a and b in T, that is, the unique node on the path P_{ab} that is ancestor of both a and b. We define $g_\vee(a, b)$ as the unique node on the path P_{ab} such that the distance between a and $g_\vee(a, b)$ is the same as the distance between b and $g_\wedge(a, b)$.

A language admitting $\langle g_\wedge, g_\vee \rangle$ as a multimorphism has been called *weakly tree-submodular*, since it has been shown that the property of strong tree-submodularity implies weak tree-submodularity [185]. The tractability of weakly tree-submodular languages on chains[3] and forks[4] has also been shown in [185].

---

[2]It is easy to show that an equivalent definition of *k*-submodularity is a generalisation of submodularity and bisubmodularity in terms of two *k*-tuples of pairwise disjoint subsets of a given finite set.

[3]A chain is a binary tree in which all nodes except leaves have exactly one child.

[4]A fork is a binary tree in which all nodes except leaves and one special node have exactly one child. The special node has exactly two children.

The tractability of strongly tree-submodular languages defined similarly on general (not necessarily binary) trees has recently been solved [264] and is discussed in Chap. 8. (Note that a special case is a tree on $k + 1$ vertices, where $k > 2$, consisting of a root node with k children. This corresponds precisely to bisubmodular languages on domains of size $k + 1$, that is, k-submodular languages from Example 6.5.) Moreover, the tractability of weakly tree-submodular languages defined on arbitrary trees has also recently been solved [264] and is discussed in Chap. 8.

Example 6.10 (1-defect) Let b and c be two distinct elements of D and let $(D; <)$ be a partial order which relates all pairs of elements except for b and c. We call $\langle f, g \rangle$, where $f, g : D^2 \to D$ are two binary operations, a 1-*defect* if f and g are both commutative and satisfy the following conditions:

- If $\{x, y\} \neq \{b, c\}$, then $f(x, y) = \text{Min}(x, y)$ and $g(x, y) = \text{Max}(x, y)$.
- If $\{x, y\} = \{b, c\}$, then $\{f(x, y), g(x, y)\} \cap \{x, y\} = \emptyset$, and $f(x, y) < g(x, y)$.

The tractability of languages that admit a 1-defect multimorphism has recently been shown in [174]. This result generalises the tractability result for weakly tree-submodular languages on chains and forks described in Example 6.9, but is incomparable with the tractability result for strongly tree-submodular languages on binary trees.

Next we present several examples of tractable languages defined by ternary multimorphisms.

Example 6.11 (Majority) A ternary operation $f : D^3 \to D$ is called a majority operation if $f(x, x, y) = f(x, y, x) = f(y, x, x) = x$ for all $x, y \in D$.

Let $\mathbf{g} = \langle g_1, g_2, g_3 \rangle$ be a triple of ternary operations such that g_1, g_2, and g_3 are all majority operations. Let $\phi : D^m \to \overline{\mathbb{Q}}_{\geq 0}$ be an m-ary cost function that admits $\mathbf{g}$ as a multimorphism. It follows from Definition 1.8 that for all $\mathbf{x}, \mathbf{y} \in D^m$, $3\phi(\mathbf{x}) \leq \phi(\mathbf{x}) + \phi(\mathbf{x}) + \phi(\mathbf{y})$ and $3\phi(\mathbf{y}) \leq \phi(\mathbf{y}) + \phi(\mathbf{y}) + \phi(\mathbf{x})$. Therefore, if both $\phi(\mathbf{x})$ and $\phi(\mathbf{y})$ are finite, then we have $\phi(\mathbf{x}) \leq \phi(\mathbf{y})$ and $\phi(\mathbf{y}) \leq \phi(\mathbf{x})$, and hence $\phi(\mathbf{x}) = \phi(\mathbf{y})$. In other words, the range of ϕ is $\{c, \infty\}$, for some finite $c \in \mathbb{Q}$, and hence ϕ is essentially crisp.

Let Γ_{Mjty} be the set of cost functions improved by a triple $\mathbf{g} = \langle g_1, g_2, g_3 \rangle$ of ternary majority operations $g_i : D^3 \to D$, $1 \leq i \leq 3$. The tractability of Γ_{Mjty} has been shown by [67], using the earlier result that CSPs closed under a majority polymorphism are tractable [165].

Example 6.12 (Minority) A ternary operation $f : D^3 \to D$ is called a minority operation if $f(x, x, y) = f(x, y, x) = f(y, x, x) = y$ for all $x, y \in D$. Let Γ_{Mnty} be the set of cost functions improved by a triple $\mathbf{g} = \langle g_1, g_2, g_3 \rangle$ of ternary minority operations $g_i : D^3 \to D$, $1 \leq i \leq 3$. A similar argument to the one in Example 6.11 shows that the cost functions in Γ_{Mnty} are essentially crisp. The tractability of Γ_{Mnty}

has been shown in [67], using the result that CSPs closed under a Mal'tsev poly-morphism[5] are tractable [92].

Example 6.13 (Majority and Minority)   Let $\mathbf{g} = \langle g_1, g_2, g_3 \rangle$ be three ternary oper-ations such that g_1 and g_2 are majority operations, and g_3 is a minority operation. Let Γ_{MJN} be the set of cost functions improved by $\mathbf{g}$. The tractability of Γ_{MJN} has been shown in [188], generalising an earlier tractability result for a specific $\mathbf{g}$ of this form from [67].

Finally, we present some tractable languages characterised by multimorphisms of arbitrary arities.

Example 6.14 (*k*-edge)   A $(k + 1)$-ary operation $f : D^{k+1} \to D$ is called a *k-edge* [155] if the following equalities hold:

$$f(x, x, y, y, y, \ldots, y, y) = y,$$

$$f(x, y, x, y, y, \ldots, y, y) = y,$$

$$f(y, y, y, x, y, \ldots, y, y) = y,$$

$$f(y, y, y, y, x, \ldots, y, y) = y,$$

$$\vdots$$

$$f(y, y, y, y, y, \ldots, x, y) = y,$$

$$f(y, y, y, y, y, \ldots, y, x) = y.$$

Special cases of $(k + 1)$-edge operations include Mal'tsev operations [40], near-unanimity operations [162], and generalised majority-minority operations [92]. Let Γ be a language that admits $\langle f, \ldots, f \rangle$ as a multimorphism for some $(k + 1)$-edge f. Applying an argument similar to the one in Example 6.11, it is easy to see that Γ is essentially crisp, and thus tractable by the result for crisp languages with a *k*-edge polymorphism [155] (cf. Remark 1.8).

6.3 Boolean Languages

Having seen in Sect. 6.2 several tractable languages characterised by multimor-phisms, we are now able to formulate a dichotomy classification of Boolean lan-guages given in [67].

Theorem 6.1 (Classification of Boolean languages)   *An arbitrary valued constraint language on* $D = \{0, 1\}$ *is tractable if it admits at least one of the following eight multimorphisms. Otherwise it is intractable.*

---

[5] A ternary operation $f : D^3 \to D$ is called Mal'tsev if $f(x, y, y) = f(y, y, x) = x$ for all $x, y \in D$.

1. $\langle g_0 \rangle$,
2. $\langle g_1 \rangle$,
3. $\langle \mathrm{Min}, \mathrm{Min} \rangle$,
4. $\langle \mathrm{Max}, \mathrm{Max} \rangle$,
5. $\langle \mathrm{Min}, \mathrm{Max} \rangle$,
6. $\langle \mathrm{Mjrty}, \mathrm{Mjrty}, \mathrm{Mjrty} \rangle$,
7. $\langle \mathrm{Mnrty}, \mathrm{Mnrty}, \mathrm{Mnrty} \rangle$,
8. $\langle \mathrm{Mjrty}, \mathrm{Mjrty}, \mathrm{Mnrty} \rangle$.

The hardness part of this classification also follows from the work of Creed and Živný [85] (more details in Sect. 2.6).

For finite-valued languages, the classification simplifies as follows [67]:

Corollary 6.1 (of Theorem 6.1) *A finite-valued constraint language on $D = \{0, 1\}$ is tractable if it admits at least one of the following three multimorphisms. Otherwise it is intractable.*

1. $\langle g_0 \rangle$,
2. $\langle g_1 \rangle$,
3. $\langle \mathrm{Min}, \mathrm{Max} \rangle$.

Using so-called cores [154, 264], which eliminate domain values not used in optimal solutions, the first two cases can be easily eliminated, thus giving:

Corollary 6.2 (of Theorem 6.1) *A core finite-valued constraint language on $D = \{0, 1\}$ is tractable if, and only if, it admits $\langle \mathrm{Min}, \mathrm{Max} \rangle$ as a multimorphism.*

This result has recently been generalised [154].

Theorem 6.2 ([154]) *Let $D = \{0, 1, 2\}$ and let Γ be a core finite-valued constraint language on D. Then Γ is tractable if, and only if, either*

- *Γ admits $\{\langle 1, \mathrm{Min} \rangle, \langle 1, \mathrm{Max} \rangle\}$ as a fractional polymorphism (with respect to some total order on D), or*
- *Γ admits $\{\langle 1, \mathrm{Min}_0 \rangle, \langle \alpha, \mathrm{Max}_0 \rangle, \langle 1 - \alpha, \mathrm{Max}_1 \rangle\}$ as a fractional polymorphism (that is, Γ is α-bisubmodular) for some real $0 < \alpha \leq 1$, up to a renaming of the domain.*

6.4 Summary

We have given a comprehensive list of known tractable languages. As we have seen, all known tractable languages can be characterised by simple fractional polymorphisms.

6.4.1 Related Work

The complexity of crisp constraint language has received a lot of attention in the literature; see [60] for more details. However, there is still no complete complexity classification of crisp languages, although many partial results have been obtained over the past 15 years. In particular, it has been shown that the tractable cases fall into two broad groups, both of which are characterised by specific algebraic conditions (polymorphisms). The first of these are the problems that can be solved by some form of local consistency [11], and the second are the problems that have a polynomial-sized generating set. The latter are characterised by k-edge polymorphisms from Example 6.14 [155]. Example 6.14 has shown that valued constraint languages admitting a k-edge multimorphism are in fact essentially crisp languages. Hence only the first condition gives rise to interesting tractable valued constraint languages. Local consistency techniques have been generalised to the VCSP, but their power has not been fully understood until recently [75]. Some recent progress on the power of soft arc consistency techniques is discussed in Chap. 8.

Raghavendra has shown in his Ph.D. thesis [240], assuming the unique games conjecture (UGC) [180], that any finite-valued VCSP can be approximated by the basic SDP (semidefinite programming) relaxation. Moreover, again under the assumption of UGC, if a VCSP is tractable then the basic SDP solves it. Unfortunately, this result does not tell us which languages are tractable.

6.4.2 Open Problems

The ultimate goal is to: (i) identify *all* tractable languages; and (ii) establish, if possible, a dichotomy theorem for languages defined on finite domains of arbitrary size, thus generalising the classification of Boolean languages presented in Theorem 6.1. As the first step, the classification of general-valued languages on three-element sets should be established. As we have seen in Theorem 6.2, finite-valued languages on three-element sets have already been classified [154], and also crisp languages on three-element sets have already been classified [39].

Chapter 7
Conservative Languages

Nothing can come of nothing.
William Shakespeare

7.1 Introduction

This chapter presents recent results on the complexity of so-called *conservative* VCSPs. These are VCSPs including all unary cost functions. The complexity classification of conservative VCSPs was obtained by Kolmogorov and Živný [188] (see [187] for the full version). This chapter reports results from their paper.

7.2 Conservative Languages

A valued constraint language Γ is called *conservative* if Γ includes all unary cost functions (cf. Examples 1.21 and 1.25). Kolmogorov and Živný have shown that this condition is polynomial-time equivalent to the following (weaker) requirement [187, Sect. 4].

Definition 7.1 A language Γ is *conservative* if Γ includes all $\{0, 1\}$-valued unary cost functions.

In order to formulate the complexity classification of conservative languages, we will need some notation.

Let Γ be a fixed language on D. Let $P = \{(a, b) \mid a, b \in D, a \neq b\}$. Let $M \subseteq P$ be arbitrary.

Let $\langle f, g \rangle$ be two binary operations. We call $\langle f, g \rangle$ a symmetric tournament pair (STP) on M (cf. Example 6.8) if f and g are conservative on $P \cup \{\{a\} \mid a \in D\}$ and commutative on M. Let $\langle \text{Mjrty}_1, \text{Mjrty}_2, \text{Mnrty}_3 \rangle$ be three ternary operations. We call $\langle \text{Mjrty}_1, \text{Mjrty}_2, \text{Mnrty}_3 \rangle$ an MJN on M (cf. Example 6.13) if operations $\text{Mjrty}_1, \text{Mjrty}_2, \text{Mnrty}_3$ are conservative and for each triple $\langle a, b, c \rangle \in D^3$ with $\{a, b, c\} = \{x, y\} \in M$ the operations $\text{Mjrty}_1(a, b, c)$, $\text{Mjrty}_2(a, b, c)$ return the unique majority element from a, b, c (which occurs twice) and $\text{Mnrty}_3(a, b, c)$ returns the remaining minority element.

S. Živný, *The Complexity of Valued Constraint Satisfaction Problems*,
Cognitive Technologies, DOI 10.1007/978-3-642-33974-5_7,
© Springer-Verlag Berlin Heidelberg 2012

Given a language Γ, we say that Γ admits *complementary STP and MJN* multimorphisms if there are a set $M \subseteq P$, binary operations $\langle f, g \rangle$, and ternary operations $\langle \mathrm{Mjrty}_1, \mathrm{Mjrty}_2, \mathrm{Mnrty}_3 \rangle$ such that $\langle f, g \rangle$ is an STP on M and $\langle \mathrm{Mjrty}_1, \mathrm{Mjrty}_2, \mathrm{Mnrty}_3 \rangle$ is an MJN on $\overline{M}$, where $\overline{M} = P \setminus M$.

Generalising the tractability results described in Examples 6.8 and 6.13, Kolmogorov and Živný have shown that languages admitting complementary STP and MJN multimorphisms are tractable [188]. Moreover, for general-valued conservative languages, it has been shown that this class is the *only* tractable case, as the following result from [188] indicates.

Theorem 7.1 (General-valued languages) *A general-valued conservative language is tractable if it admits complementary STP and MJN multimorphisms. Otherwise it is intractable.*

In the finite-valued case, the more restricted class from Example 6.8 is the only tractable case, as the following result from [188] indicates.

Theorem 7.2 (Finite-valued languages) *A finite-valued conservative language is tractable if it admits a symmetric tournament pair multimorphism. Otherwise it is intractable.*

In fact, these classification results also hold for (the slightly more general) fixed-value languages (cf. Example 1.16).

7.3 Graph Technique

In this section we will briefly discuss the techniques used to prove Theorems 7.1 and 7.2. The key idea is to relate the complexity of a conservative language Γ to properties of a certain graph G_Γ associated with Γ.

Given a conservative language Γ, let $G_\Gamma = (P, E)$ be the graph with the set of nodes $P = \{(a, b) \mid a, b \in D, a \neq b\}$ and the set of edges E defined as follows: there is an edge between $(a, b) \in P$ and $(a', b') \in P$ if, and only if, there exists a binary cost function $\phi \in \langle \Gamma \rangle$ such that

$$\phi(a, a') + \phi(b, b') > \phi(a, b') + \phi(b, a'). \tag{7.1}$$

The idea is that ϕ is not submodular with respect to an order $<$ on D in which $a < b$ and $a' < b'$, or $a > b$ and $a' > b'$.

Note that G_Γ may have self-loops.

We say that edge $\{(a, b), (a', b')\} \in E$ is *soft* if there exists a binary $\phi \in \langle \Gamma \rangle$ satisfying (7.1) such that at least one of the costs $\phi(a, a')$, $\phi(b, b')$ is finite. Edges in E that are not soft are called *hard*. For node $p = (a, b) \in P$ we denote $\bar{p} = (b, a) \in P$.

We define $M \subseteq P$ to be the set of vertices $(a, b) \in P$ without self-loops, and $\overline{M} = P - M$ to be the complement of M. It follows from the definition that the set M is *symmetric*, that is, $(a, b) \in M$ if, and only if, $(b, a) \in M$. We will write $\{a, b\} \in M$ to indicate that $(a, b) \in M$; this is consistent due to the symmetry of M. Similarly, we will write $\{a, b\} \in \overline{M}$ if $(a, b) \in \overline{M}$, and $\{a, b\} \in P$ if $(a, b) \in P$, that is, $a, b \in D$ and $a \neq b$.

Theorems 7.1 and 7.2 follow from the following three theorems.

Theorem 7.3 *Let Γ be a conservative language.*

1. *If G_Γ has a soft self-loop then Γ is intractable.*
2. *If G_Γ does not have soft self-loops then Γ admits a pair $\langle f, g \rangle$ which is an STP on M and satisfies additionally $f(a, b) = a$, $g(a, b) = b$ for $\{a, b\} \in \overline{M}$.*

Theorem 7.4 *Let Γ be a conservative language. If Γ does not admit an MJN on $\overline{M}$ then it is intractable.*

Theorem 7.5 *Suppose a language Γ admits an STP on M and an MJN on $\overline{M}$, for some choice of symmetric $M \subseteq P$. Then Γ is tractable.*

Theorems 7.3, 7.4, and 7.5 give the dichotomy result for conservative languages and thus prove Theorem 7.1: If a conservative language Γ admits an STP on M and an MJN on $\overline{M}$ for some symmetric $M \subseteq P$ then Γ is tractable by Theorem 7.5. Otherwise Γ is intractable: If G_Γ has a soft self-loop, then Γ is intractable by Theorem 7.3(1); if G_Γ does not have soft self-loops, then Γ admits an STP on M by Theorem 7.3(2); if G_Γ does not have soft self-loops and Γ does not admit an MJN on $\overline{M}$, then Γ is intractable by Theorem 7.4.

In the finite-valued case, we get a simpler tractability criterion as stated in Theorem 7.2: If a conservative finite-valued language Γ admits an STP then Γ is tractable. Otherwise Γ is intractable. The proof of Theorem 7.2 follows from Theorems 7.3 and 7.5: Consider the graph G_Γ associated with Γ. If G_Γ contains a soft self-loop, then, by Theorem 7.3(1), Γ is intractable. Suppose that G_Γ does not contain soft self-loops. As Γ is finite-valued, G_Γ cannot have hard self-loops. Therefore, $\overline{M}$ is empty and $M = P$. By Theorem 7.3(2), Γ admits an STP. The tractability then follows from Theorem 7.5.

The algorithm proving Theorem 7.5 is based on a preprocessing step that reduces a given instance to an instance admitting an STP multimorphism. The proof of Theorem 7.4 is fairly technical. The rest of this section is devoted to the ideas that lie behind the proof of Theorem 7.3, which is sufficient for the classification of finite-valued languages.

Using unary cost functions, Theorem 7.3(1) can be proved via a reduction from the Max-SAT problem with XOR clauses and the independent set problem. We focus on Theorem 7.3(2). Graph G_Γ and its properties play a crucial role in the proofs.

The following lemma can be derived from the definition of G_Γ [187, 188].

Lemma 7.1 *Graph $G_\Gamma = (P, E)$ defined above satisfies the following properties:*

(a) $\{p, q\} \in E$ implies $\{\bar{p}, \bar{q}\} \in E$, and vice versa.
(b) The induced subgraph $(M, E[M])$ does not have odd cycles.
(c) If node p is not isolated (that is, it has at least one incident edge $\{p, q\} \in E$) then $\{p, \bar{p}\} \in E$.

We now construct a pair of operations $\langle f, g \rangle$ for Γ that behaves as an STP on M and as a multi-projection (returning its two arguments in the same order) on $\overline{M}$.

Lemma 7.2 There exists an assignment $\sigma : M \rightarrow \{-1, +1\}$ such that (i) $\sigma(p) = -\sigma(q)$ for all edges $\{p, q\} \in E$, and (ii) $\sigma(p) = -\sigma(\bar{p})$ for all $p \in M$.

Proof By Lemma 7.1(b), graph $(M, E[M])$ does not have odd cycles. Therefore, by Harary's Theorem, graph $(M, E[M])$ is bipartite and there exists an assignment $\sigma : M \rightarrow \{-1, +1\}$ that satisfies property (i). Let us modify this assignment as follows: for each isolated node $p \in M$ (that is, node without incident edges) set $\sigma(p)$, $\sigma(\bar{p})$ so that $\sigma(p) = -\sigma(\bar{p}) \in \{-1, +1\}$. (Note, if p is isolated then by Lemma 7.1(a) so is $\bar{p}$.) Clearly, property (i) still holds. Property (ii) holds for each node $p \in M$ as well: if p is isolated then (ii) holds by construction; otherwise by Lemma 7.1(c) there exists edge $\{p, \bar{p}\} \in E$, and so (ii) follows from property (i). $\square$

Given an assignment σ constructed as in Lemma 7.2, we now define operations $f, g : D^2 \rightarrow D$ as follows:

- $f(a, a) = g(a, a) = a$ for $a \in D$.
- If $(a, b) \in M$ then $f(a, b)$ and $g(a, b)$ are the unique elements of D satisfying $\{f(a, b), g(a, b)\} = \{a, b\}$ and $\sigma(f(a, b), g(a, b)) = +1$.
- If $(a, b) \in \overline{M}$ then $f(a, b) = a$ and $g(a, b) = b$.

Using induction on the arity of the cost functions in $\langle \Gamma \rangle$ one can show that $\langle f, g \rangle$ is a multimorphism of $\langle \Gamma \rangle$ [188]. The main idea behind this proof comes from the proof that a k-ary finite-valued cost function f is submodular if, and only if, every binary projection of f is submodular [265]. However, this is known to not hold for general-valued cost functions (that is, cost functions taking on both finite and infinite costs) [49], and hence the actual proofs are more elaborate.

Note that G_Γ depends on the infinite set $\langle \Gamma \rangle$, but it is not necessary to actually construct G_Γ to decide whether Γ admits complementary STP and MJN multimorphisms, which can be checked in polynomial time in $|D|$ and $|\Gamma|$. (The same is true in the finite-valued case, where the existence of an STP multimorphism guarantees tractability.) Hence, the criteria given in Theorems 7.1 and 7.2 are decidable, and can be computed in polynomial time.

7.4 Summary

We have briefly introduced a recently developed graph technique for studying the computational complexity of VCSPs. Moreover, we have presented the resulting classification of conservative languages.

7.4.1 Related Work

Bulatov's work on conservative CSPs [45] (cf. Example 1.23) was the first to introduce the idea of capturing the interaction between two-element subdomains via a finite graph. A somewhat similar graph to the one presented here (but not exactly the same) has been used by Takhanov [261] for languages Γ containing crisp functions and finite unary cost functions (cf. Example 1.24).[1]

The graph technique presented in this chapter can be used to simplify previously obtained classifications of Max-CSPs on three-element domains [171] and fixed-value Max-CSPs [100], which both depend on computer-assisted search. Moreover, using the graph technique, Jonsson et al. have recently presented a complete classification of Max-CSPs on four-element domains [174],[2] thus demonstrating the usefulness of this technique beyond conservative languages.[3]

7.4.2 Open Problems

As mentioned in the previous paragraph, the graph technique discussed in this section has proved important even in the context of non-conservative languages [174]. Another example where the graph technique has proved useful is approximate counting [47, 57]. We believe that it will be used in many other contexts in the future.

---

[1]Roughly speaking, the graph structure in [261] was defined via a "min" polymorphism rather than a $\langle \text{Min}, \text{Max} \rangle$ multimorphism. The property $\{p, q\} \in E \Rightarrow \{\bar{p}, \bar{q}\} \in E$ might not hold in Takhanov's case. Also, in [261] edges were not classified as being soft or hard.

[2]The only two tractable classes over these domain sizes are those characterised by $\langle \text{Min}, \text{Max} \rangle$ multimorphisms (cf. Example 6.4) and those characterised by 1-defect multimorphisms (cf. Example 6.10).

[3]One needs to modify slightly the definition of G_Γ so that an edge between two nodes corresponds to a violation of the definition of a binary multimorphism, rather than to just the definition of a submodular multimorphism.

Chapter 8
The Power of Linear Programming

> *You don't understand anything until you learn it more than one way.*
> Marvin Minsky

8.1 Introduction

This chapter presents recent results on the power of linear programming for VC-SPs. In particular, we present an algebraic characterisation (in terms of fractional polymorphisms) of languages that are tractable via a standard linear programming relaxation. This result has recently been obtained by Thapper and Živný [264]. We also discuss algorithmic consequences of this characterisation. This chapter summarises results from [264].

8.2 Basic LP Relaxation

In order to be consistent with the notation used in [264], we now introduce a slightly different terminology for VCSPs.

A *signature* τ is a set of *function symbols* ϕ, each with an associated positive *arity*, $\mathrm{ar}(\phi)$. A *finite-valued* τ-*structure* A (previously called a finite-valued language) consists of a *domain* $D = D(A)$, together with a cost function $\phi^A : D^{\mathrm{ar}(f)} \to \mathbb{Q}_{\geq 0}$, for each function symbol $\phi \in \tau$.

Let A be a finite-valued τ-structure. An instance of $\mathsf{VCSP}(A)$ is given by a finite-valued τ-structure I. A solution to I is a function $h : D(I) \to D(A)$, its measure given by

$$\sum_{\phi \in \tau, \bar{x} \in D(I)^{\mathrm{ar}(\phi)}} \phi^I(\bar{x}) \phi^A\big(h(\bar{x})\big).$$

The goal is to find a solution of minimum measure. This measure will be denoted by $\mathrm{Opt}_A(I)$.

It should be clear that this definition is equivalent to the original definition of VCSPs (cf. Definition 1.1 in Chap. 1). Cost functions from the (finite-valued τ-structure) instance I correspond to (weighted) scopes.

S. Živný, *The Complexity of Valued Constraint Satisfaction Problems*,
Cognitive Technologies, DOI 10.1007/978-3-642-33974-5_8,
© Springer-Verlag Berlin Heidelberg 2012

For an m-tuple $\bar{t}$, we denote by $\{\bar{t}\}$ the set of elements in $\bar{t}$. Furthermore, we denote by $[\bar{t}]$ the multiset of elements in $\bar{t}$.

Let I and A be valued structures over a common finite signature τ. Let $X = D(I)$ and $D = D(A)$. The *basic LP relaxation* (BLP) (sometimes also called the *standard*, or *canonical LP relaxation*) has variables $\lambda_{\phi,\bar{x},\sigma}$ for $\phi \in \tau$, $\bar{x} \in X^{\text{ar}(\phi)}$, $\sigma : \{\bar{x}\} \to D$, and variables $\mu_x(a)$ for $x \in X$, $a \in D$.

$$\min \quad \sum_{\phi,\bar{x}} \sum_{\sigma:\{\bar{x}\} \to D} \phi^I(\bar{x}) \phi^A\big(\sigma(\bar{x})\big) \lambda_{\phi,\bar{x},\sigma},$$

$$\text{s.t.} \quad \sum_{\sigma:\sigma(x)=a} \lambda_{\phi,\bar{x},\sigma} = \mu_x(a) \quad \forall \phi \in \tau, \ \bar{x} \in X^{\text{ar}(\phi)}, \ x \in \{\bar{x}\}, \ a \in D,$$

$$\sum_{a \in D} \mu_x(a) = 1 \quad \forall x \in X, \tag{8.1}$$

$$0 \leq \lambda, \quad \mu \leq 1.$$

For any fixed A, BLP is polynomial in the size of a given VCSP(A) instance. Let IP be the program obtained from (8.1) with the requirement that all variables take values in the range $\{0, 1\}$ rather than $[0, 1]$. This is an integer programming formulation of the original VCSP instance. The interpretation of the variables in IP is as follows: $\mu_x(a) = 1$ if, and only if, variable x is assigned value a; $\lambda_{\phi,\bar{x},\sigma} = 1$ if, and only if, constraint ϕ on scope $\bar{x}$ is assigned tuple $\sigma(\bar{x})$. LP (8.1) is now a relaxation of IP and the question of whether (8.1) solves a given VCSP instance I is the question of whether IP has a zero integrality gap.

In order to state the main theorem, we will need some more notation.

Let A and B be valued structures over the same signature τ. Let B^A denote the set of all functions from $D(A)$ to $D(B)$. A *fractional homomorphism* from A to B is a function $\omega : B^A \to \mathbb{Q}_{\geq 0}$, with $\sum_{g \in B^A} \omega(g) = 1$, such that for every function symbol $\phi \in \tau$ and tuple $\bar{a} \in D(A)^{\text{ar}(\phi)}$, it holds that

$$\sum_{g \in B^A} \omega(g) \phi^B\big(g(\bar{a})\big) \leq \phi^A(\bar{a}),$$

where the functions g are applied componentwise. We write $A \to_f B$ to indicate the existence of a fractional homomorphism.

Let A be a valued τ-structure, $D = D(A)$, and let $m \geq 1$. We define the *multiset-structure*[1] $P^m(A)$ as the valued structure with domain $\left(\binom{D}{m}\right)$, where $\left(\binom{D}{m}\right)$ denotes the multisets of elements from D of size m, and for every k-ary function symbol $\phi \in \tau$, and $\alpha_1, \ldots, \alpha_k \in \left(\binom{D}{m}\right)$,

$$\phi^{P^m(A)}(\alpha_1, \ldots, \alpha_k) = \frac{1}{m} \min_{\bar{t}_i \in D^m : [\bar{t}_i] = \alpha_i} \sum_{i=1}^{m} \phi^A\big(\bar{t}_1[i], \ldots, \bar{t}_k[i]\big).$$

---

[1] A similar structure for $\{0, \infty\}$-valued languages was introduced in [197].

For a fractional polymorphism $\mathcal{F} = \langle\langle r_1, f_1\rangle, \ldots, \langle r_n, f_n\rangle\rangle$, we define supp($\mathcal{F}$) = $\{f_1, \ldots, f_n\}$.

Let S_m be the symmetric group on $\{1, \ldots, m\}$. An m-ary operation f is *symmetric* if for every permutation $\pi \in S_m$, we have $f(x_1, \ldots, x_m) = f(x_{\pi(1)}, \ldots, x_{\pi(m)})$. A *symmetric fractional polymorphism* $\mathcal{F}$ is a fractional polymorphism such that if $g \in$ supp($\mathcal{F}$), then g is symmetric.

We are now ready to state the main result.

Theorem 8.1 ([264]) *Let A be a finite-valued structure over a finite signature. The following are equivalent:*

1. *BLP solves VCSP(A).*
2. *For every $m > 1$, $P^m(A) \to_f A$.*
3. *For every $m > 1$, A has an m-ary symmetric fractional polymorphism.*
4. *For every $n > 1$, A has a fractional polymorphism $\mathcal{F}_n$ such that supp($\mathcal{F}_n$) generates an n-ary symmetric operation.*

The proof of Theorem 8.1 uses, apart from other techniques, Farkas' Lemma. The most surprising is the fourth statement of Theorem 8.1, which requires a rather technical proof.

The reader might have noticed that this section mentions only finite-valued τ-structures (languages). In fact, the same result holds true for general-valued τ-structures, where the BLP algorithm is augmented with a standard preprocessing step known as arc consistency [118].[2] Moreover, the BLP algorithm, as described above, gives the cost of an optimal solution, but does not give an optimal solution. This can be done using the idea of self-reduction. We refer the reader to [264] for more details.

8.3 Algorithmic Consequences

Using Theorem 8.1, we will show tractability of various languages, including languages previously unknown to be tractable. In particular, we show that all known finite-valued tractable languages are solved by BLP.

Since any semi-lattice operation[3] generates symmetric operations of all arities, we get:

Corollary 8.1 (of Theorem 8.1) *Let A be a valued structure with a binary multimorphism $\langle g_1, g_2\rangle$ where either g_1 or g_2 is a semi-lattice operation. Then A is tractable.*

---

[2]The algorithm from [118] is sometimes called the generalised arc consistency algorithm to emphasise the fact that it works for CSPs of arbitrary arities, and not only for binary CSPs [207].

[3]A semi-lattice operation is associative, commutative, and idempotent.

We now give examples of valued structures (that is, valued constraint languages) defined by such binary multimorphisms.

Example 8.1 Let $(D; \wedge, \vee)$ be an *arbitrary* lattice on D. Assume that a valued structure A has the multimorphism $\langle \wedge, \vee \rangle$. Then VCSP($A$) is tractable. The tractability of A was previously known only for distributive lattices [159, 257] (cf. Example 6.4) and (finite-valued) diamonds [192]; see also [191].

Example 8.2 Recall from Example 6.8 that a pair of operations $\langle g_1, g_2 \rangle$ is called a symmetric tournament pair (STP) if both g_1 and g_2 are commutative, are conservative ($g_1(x, y) \in \{x, y\}$ and $g_2(x, y) \in \{x, y\}$ for all $x, y \in D$), and $g_1(x, y) \neq g_2(x, y)$ for all $x, y \in D$. Let A be a finite-valued structure with an STP multimorphism $\langle g_1, g_2 \rangle$. It is known that if a finite-valued structure admits an STP multimorphism, it also admits a submodularity multimorphism. This result is implicitly contained in [66].[4] Therefore, BLP solves any instance from VCSP(A).

Example 8.3 Assume that a valued structure A is bisubmodular [122]. This means that $D = \{0, 1, 2\}$ and A has a multimorphism $\langle \text{Min}_0, \text{Max}_0 \rangle$ (cf. Example 6.5). Since Min_0 is a semi-lattice operation, A is tractable. The tractability of (finite-valued) A was previously known only using a general algorithm for bisubmodular functions given by an oracle [122, 217].

Example 8.4 As in Example 8.3, α-bisubmodular structures [154] (cf. Example 6.6) are solvable by BLP for all $0 < \alpha \leq 1$.

Example 8.5 Assume that a valued structure A is weakly tree-submodular on an *arbitrary* tree [185] (cf. Example 6.9). The meet (which is defined to be the highest common ancestor) is again a semi-lattice operation. The same holds for strongly tree-submodular structures since strong tree-submodularity implies weak tree-submodularity [185].

The tractability of weakly tree-submodular valued structures was previously known only for chains and forks [185]. The tractability of strongly tree-submodular valued structures was previously known only for binary trees [185].

Example 8.6 Note that the previous example applies to *all* trees, not just binary ones. In particular, it applies to the tree consisting of one root with k children. This is equivalent to structures with $D = \{0, 1, \ldots, k\}$ and the multimorphism $\langle \text{Min}_0, \text{Max}_0 \rangle$ from Example 8.3. It is a natural generalisation of submodular ($k = 1$) and bisubmodular ($k = 2$) functions, known as k-submodular functions [153]. The tractability of k-submodular valued structures for $k > 2$ was previously open.

---

[4]Namely, the STP might contain cycles, but [66, Lemma 7.15] tells us that on cycles we have, in the finite-valued case, only unary cost functions. It follows that the cost functions admitting the STP must be submodular with respect to some total order.

Example 8.7 Recall the definition of a 1-defect multimorphism $\langle g_1, g_2 \rangle$ from Example 6.10. The tractability of valued structures that have a 1-defect multimorphism has been shown in [174]. It can be shown that either g_1 or g_2 generates symmetric operations of all arities [264], and thus BLP solves any instance admitting a 1-defect multimorphism.

8.4 Summary

We have briefly surveyed the basic linear programming relaxation and presented its power in terms of fractional polymorphisms together with its algorithmic consequences.

8.4.1 Related Work

Some other approaches to using linear programming for VCSPs have been studied in [73, 75] and in the context of computer vision [268, 269]. In particular, as the dual of the so-called *optimal soft arc consistency* (OSAC) [75] is a tighter relaxation than BLP (see [264] for details), OSAC solves all problems solved by BLP.

Semidefinite programming (SDP) is a powerful tool for solving optimisation problems. SDP is known to be strictly more powerful than LP for some problems. As mentioned in Chap. 6, Raghavendra has shown, assuming the unique games conjecture (UGC) [180], that if a finite-valued VCSP is tractable, then the basic SDP relaxation solves it [240].

8.4.2 Open Problems

None of the conditions in Theorem 8.1 is immediately decidable, let alone checkable in polynomial time in the size of the valued structure A. It would be interesting to know whether there is another equivalent characterisation that gives a decidable criterion.

BLP solves all known finite-valued tractable languages. Is it possible that in fact BLP solves *all* tractable languages?

Chapter 9
Hybrid Tractability

> *Never had any mathematical conversations with anybody,*
> *because there was nobody else in my field.*
> Alonzo Church

9.1 Introduction

This chapter is devoted to recent results on hybrid tractability of VCSPs, that is, reasons for tractability that do not follow from the restriction on the cost functions (such as submodularity) or on the structure of the instance (such as bounded treewidth).

Hybrid tractability is a fairly recent research direction, initiated by the work of Cooper and Živný. This chapter summarises results from [80, 82].

9.2 Tractable Triangles

In this section, we will only deal with *binary* VCSPs, where every constraint is of arity at most 2. However, in this chapter, the domain of values for the variables is not fixed and is part of the input. (This is in contrast with the rest of this book, where we assume that the domain is fixed.)

A (binary) VCSP instance is given by n variables $v_1, \ldots, v_n$ over finite domains $D_1, \ldots, D_n$ of values. (That is, we allow different domains for different variables.) Without loss of generality, we can assume that any instance contains constraints of all possible scopes, that is, n unary constraints and $\binom{n}{2}$ binary constraints. We denote the cost function associated with the unary constraint with the scope $\langle v_i \rangle$ by c_i and the cost function associated with the binary constraint with the scope $\langle v_i, v_j \rangle$ by c_{ij}. The absence of any constraint on variable v_i (or between variables v_i, v_j) is modelled by a cost function c_i (or c_{ij}) that is uniformly zero.

Definition 9.1 (Triangle) In a VCSP instance, a *triangle* is a set of assignments $\{\langle v_i, a \rangle, \langle v_j, b \rangle, \langle v_k, c \rangle\}$, where v_i, v_j, v_k are distinct variables and $a \in D_i$, $b \in D_j$, $c \in D_k$ are domain values. The multiset of costs in such a triangle is

S. Živný, *The Complexity of Valued Constraint Satisfaction Problems*,
Cognitive Technologies, DOI 10.1007/978-3-642-33974-5_9,
© Springer-Verlag Berlin Heidelberg 2012

$\{c_{ij}(a,b), c_{ik}(a,c), c_{jk}(b,c)\}.$[1] A *triple of costs* will always refer to a multiset of binary costs in a triangle.

Definition 9.2 A triangle $\{\langle v_i, a\rangle, \langle v_j, b\rangle, \langle v_k, c\rangle\}$, where $a \in D_i, b \in D_j, c \in D_k$, satisfies the *joint-winner property* (JWP) if either all three $c_{ij}(a,b)$, $c_{ik}(a,c)$, $c_{jk}(b,c)$ are the same, or two of them are equal and the third one is larger. A VCSP instance satisfies the joint-winner property if every triangle satisfies the joint-winner property.

Example 9.1 ((Soft) All Different) One of the most commonly used global constraints is the ALLDIFFERENT constraint [242]. Petit et al. introduced a soft version of the ALLDIFFERENT constraint, SOFTALLDIFF [231]. They proposed two types of costs to be minimised: graph- and variable-based costs. The former counts the number of equalities, whilst the latter counts the minimum number of variables that need to change value in order to satisfy the ALLDIFFERENT constraint. The algorithms for filtering these constraints, introduced in the same paper, were then improved by van Hoeve et al. [152].

It is easy to check that the graph-based variant of the SOFTALLDIFF constraint satisfies the joint-winner property. In this case, for every i and j, the cost function c_{ij} is defined by $c_{ij}(a,b) = 1$ if $a = b$, and $c_{ij}(a,b) = 0$ otherwise. Take any three variables v_i, v_j, v_k and $a \in D_i, b \in D_j, c \in D_k$. If $c_{ij}(a,b) = c_{jk}(b,c) = c_{ik}(a,c)$ (which means that the domain values a, b, c are all equal or all different), then the joint-winner property is satisfied trivially. If only one of the costs is 1, then the joint-winner property is satisfied as well. Observe that due to the transitivity of equality it cannot happen that only one of the costs is 0.

In order to code the variable-based SOFTALLDIFF constraint as a binary VCSP $\mathcal{P}$, we can create n variables v_i' with domains $D_i \times \{1, 2\}$. The assignment $v_i' = (a, 1)$ means that v_i is the first variable of the original VCSP to be assigned the value a, whereas $v_i' = (a, 2)$ means that v_i is assigned a but it is not the first such variable. In $\mathcal{P}$ there is a crisp constraint that disallows $v_i' = v_j' = (a, 1)$ (for any value $a \in D_i \cap D_j$) for each pair of variables $i < j$ together with a soft unary constraint $c_i(a, k) = k - 1$ (for $k = 1, 2$) for each $i \in \{1, \ldots, n\}$. Hence at most one variable can be the first to be assigned a, and each assignment of the value a to a variable (apart from the first) incurs a cost of 1. Again due to the transitivity of equality, it cannot happen that only one of the binary costs is 0, from which it follows immediately that the joint-winner property is satisfied in $\mathcal{P}$.

One of the first results on hybrid tractability of VCSPs is the following:

Theorem 9.1 ([80]) *The class of VCSPs satisfying JWP is tractable.*

---

[1] Note that the unary costs $c_i(a), c_j(b), c_k(c)$ are not considered here.

Moreover, Cooper and Živný have also shown [80] that the class defined by the joint-winner property is maximal—allowing a single extra triple of costs that violates the joint-winner property renders the class NP-hard.

Theorem 9.2 ([80]) *Let* $\alpha < \beta \leq \gamma$, *where* $\alpha \in \mathbb{Q}_{\geq 0}$ *and* $\beta, \gamma \in \overline{\mathbb{Q}}_{\geq 0}$, *be a multiset of costs that do not satisfy the joint-winner property. The class of instances where the costs in each triangle either satisfy the joint-winner property or are* $\{\alpha, \beta, \gamma\}$ *is NP-hard, even for Boolean Max-CSPs, CSPs over size-3 domains or Boolean finite-valued VCSPs.*

In this section we consider a broader question, whether allowing any fixed set S of triples of costs in triangles, where S does not necessarily include all triples allowed by the JWP, defines a tractable class of VCSP instances.

In the case of CSP, there are only four possible multisets of costs ($\{0, 0, 0\}$, $\{0, 0, \infty\}$, $\{0, \infty, \infty\}$, $\{\infty, \infty, \infty\}$) and it is possible to study all 16 subsets S of this set. But, given an infinite set of possible costs, such as $\mathbb{Q}_{\geq 0}$ or $\overline{\mathbb{Q}}_{\geq 0}$, there is an infinite number of sets S of triples of costs. Obviously, we cannot consider all such sets. Therefore, we only consider cases defined by the total order $<$ on the set of costs, Ω, corresponding to a partition of the set of all possible triples of costs into a small number of types of triples.

Definition 9.3 Let $\mathfrak{D}$ denote the set of all possible cost types under consideration. Let Ω be a fixed set of allowed costs. For any $S \subseteq \mathfrak{D}$, we denote by $\mathcal{A}_\Omega(S)$ ($\mathcal{A}$ for allowed) the set of binary VCSP instances whose costs lie in Ω and where the triples of costs in all triangles belong to S.

Our goal is to classify the complexity of $\mathcal{A}_\Omega(S)$ for every $S \subseteq \mathfrak{D}$. The problem $\mathcal{A}_\Omega(S)$ is considered *tractable* if there is a polynomial-time algorithm to solve it and *intractable* if it is NP-hard.

Proposition 9.1 *Let* Ω *be an arbitrary set of costs and* S *be a set of cost types.*

1. *If* $\mathcal{A}_\Omega(S)$ *is tractable and* $S' \subseteq S$, *then* $\mathcal{A}_\Omega(S')$ *is tractable.*
2. *If* $\mathcal{A}_\Omega(S)$ *is intractable and* $S' \supseteq S$, *then* $\mathcal{A}_\Omega(S')$ *is intractable.*

Remark 9.1 We implicitly allow all unary cost functions. In fact, all tractability results reported in this section work with unary cost functions, and all NP-hardness results do not require any unary cost functions.

Remark 9.2 We consider problems with unbounded domains, that is, the domain sizes are part of the input. However, all NP-hardness results are obtained for problems with a fixed domain size.[2] In the case of CSPs, we need domains of size 3 to

---

[2]In other words, the considered problems are not fixed-parameter tractable [102] in the domain size.

prove NP-hardness, and in all other cases domains of size 2 are sufficient to prove NP-hardness. Since binary CSPs are known to be tractable on Boolean domains, and any VCSP is trivially tractable over domains of size 1, all NP-hardness results are tight.

9.2.1 CSPs

In this section, we will focus on the set of possible costs $\Omega = \{0, \infty\}$, that is, constraint satisfaction problems (CSPs). We consider the four following types of triples of costs:

Symbol	Costs
$<$	$\{0, 0, \infty\}$
$>$	$\{0, \infty, \infty\}$
0	$\{0, 0, 0\}$
∞	$\{\infty, \infty, \infty\}$

The set of possible cost types is thus $\mathfrak{D} = \{<, >, 0, \infty\}$. Indeed, these four cost types correspond precisely to the four possible multisets of costs: $\{0, 0, 0\}$, $\{0, 0, \infty\}$, $\{0, \infty, \infty\}$, and $\{\infty, \infty, \infty\}$. The dichotomy presented in this section therefore represents a complete characterisation of the complexity of CSPs defined by placing restrictions on triples of costs in triangles.

As $\mathcal{A}_{\{0,\infty\}}(\mathfrak{D})$ allows all binary CSPs, $\mathcal{A}_{\{0,\infty\}}(\mathfrak{D})$ is intractable [229] unless the domain is of size at most 2, which is equivalent to 2-SAT, and a well-known tractable class [252].

The joint-winner property for CSPs gives

Corollary 9.1 (of Theorem 9.1) $\mathcal{A}_{\{0,\infty\}}(\{<, 0, \infty\})$ *is tractable.*

Cooper and Živný have shown that the following two classes are tractable (for rather trivial reasons): $\mathcal{A}_{\{0,\infty\}}(\{>, 0, \infty\})$, and $\mathcal{A}_{\{0,\infty\}}(\{<, >, \infty\})$. In fact, they have established the following complexity classification, which is depicted in Fig. 9.1: white nodes represent tractable cases and shaded nodes represent intractable cases.

Theorem 9.3 ([82]) *For* $|D| \geq 3$, *a class of binary CSP instances defined by* $\mathcal{A}_{\{0,\infty\}}(S)$, *where* $S \subseteq \{<, >, 0, \infty\}$, *is intractable if, and only if,* $\{<, >, 0\} \subseteq S$.

A simple way to convert classical CSPs into an optimisation problem is to allow soft unary constraints. This framework includes well-studied problems such as MAX-ONES over Boolean domains [87, 179] (cf. Example 1.4) and non-Boolean domains [173], Max-Solution [175] (cf. Example 1.26), and Min-Cost-Hom [261] (cf. Example 1.24).

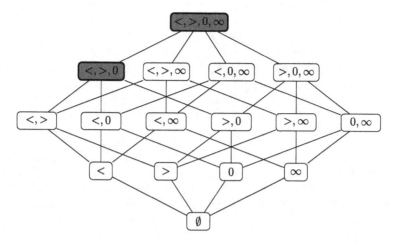

Fig. 9.1 Complexity of CSPs $\mathcal{A}_{\{0,\infty\}}(S)$, $S \subseteq \{<, >, 0, \infty\}$

It turns out that the dichotomy given in Theorem 9.3 remains valid even if soft unary constraints are allowed. In this case, the intractable cases are now intractable even for domains of size 2.

We use the notation $\mathcal{A}_{\{0,\infty\}}^{\mathbb{Q}_{\geq 0}}(S)$ to represent the set of VCSP instances with binary costs from $\{0, \infty\}$ and unary costs from $\mathbb{Q}_{\geq 0}$ whose triples of costs in triangles belong to S. In other words, we now consider VCSPs with crisp binary constraints and soft unary constraints.

Theorem 9.4 ([82]) *For $|D| \geq 2$, a class of binary CSP instances defined by $\mathcal{A}_{\{0,\infty\}}^{\mathbb{Q}_{\geq 0}}(S)$, where $S \subseteq \{<, >, 0, \infty\}$, is intractable if, and only if, $\{<, >, 0\} \subseteq S$.*

9.2.2 Max-CSPs

In this section, we will focus on the set of possible costs $\Omega = \{0, 1\}$. It is well known that the VCSP with costs in $\{0, 1\}$ is polynomial-time equivalent to unweighted Max-CSP (no repetition of constraints allowed) [248]. The four types of triples of costs we consider are:

Symbol	Costs
$<$	$\{0, 0, 1\}$
$>$	$\{0, 1, 1\}$
0	$\{0, 0, 0\}$
1	$\{1, 1, 1\}$

The set of possible cost types is then $\mathfrak{D} = \{<, >, 0, 1\}$. Again, these four costs types correspond precisely to the four possible multisets of costs: $\{0, 0, 0\}$, $\{0, 0, 1\}$,

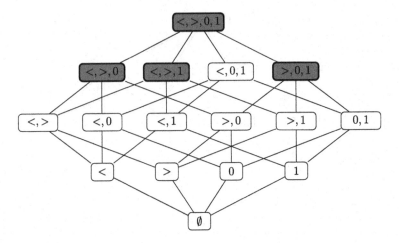

Fig. 9.2 Complexity of Max-CSPs $\mathcal{A}_{\{0,1\}}(S)$, $S \subseteq \{<, >, 0, 1\}$

$\{0, 1, 1\}$, and $\{1, 1, 1\}$. As for the CSP, the dichotomy result for Max-CSP represents a complete characterisation of the complexity of classes of instances defined by placing restrictions on triples of costs in triangles.

As $\mathcal{A}_{\{0,1\}}(\mathfrak{D})$ allows all binary Max-CSPs, $\mathcal{A}_{\{0,1\}}(\mathfrak{D})$ is intractable [125, 229] unless the domain is of size 1.

The joint-winner property [80] for Max-CSPs gives

Corollary 9.2 (of Theorem 9.1) $\mathcal{A}_{\{0,1\}}(\{<, 0, 1\})$ *is tractable.*

The following classes have been shown to be tractable by Cooper and Živný [82]: $\mathcal{A}_{\{0,1\}}(\{<, >\})$, $\mathcal{A}_{\{0,1\}}(\{>, 0\})$, and $\mathcal{A}_{\{0,1\}}(\{>, 1\})$ (via maximum matching in graphs [105]).

Moreover, Cooper and Živný have established the following complexity classification, which is depicted in Fig. 9.2: white nodes represent tractable cases and shaded nodes represent intractable cases.

Theorem 9.5 *For* $|D| \geq 2$, *a class of binary unweighted Max-CSP instances defined by* $\mathcal{A}_{\{0,1\}}(S)$, *where* $S \subseteq \{<, >, 0, 1\}$, *is intractable if, and only if, either* $\{<, >, 0\} \subseteq S$, $\{<, >, 1\} \subseteq S$, *or* $\{>, 0, 1\} \subseteq S$.

9.2.3 Finite-Valued VCSPs

In this section, we will focus on finite-valued VCSPs. In other words, we consider the set of possible costs $\Omega = \mathbb{Q}_{\geq 0}$. Since there are an infinite number of triples of costs, we consider types of triples defined by the total order on Ω. We study three different ways of partitioning the set of all triples of costs into distinct types.

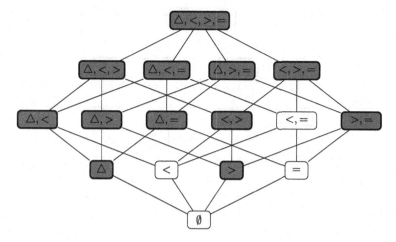

Fig. 9.3 Complexity of finite-valued VCSPs $\mathcal{A}_{\mathbb{Q}_{\geq 0}}(S)$, $S \subseteq \{\triangle, <, >, =\}$

Classification with Respect to Order   The set of possible cost types is $\mathfrak{D} = \{\triangle, <, >, =\}$, where these four types are defined in the following table:

Symbol	Costs	Remark
$\triangle$	$\{\alpha, \beta, \gamma\}$	$\alpha, \beta, \gamma \in \Omega, \alpha \neq \beta \neq \gamma \neq \alpha$
$<$	$\{\alpha, \alpha, \beta\}$	$\alpha, \beta \in \Omega, \alpha < \beta$
$>$	$\{\alpha, \alpha, \beta\}$	$\alpha, \beta \in \Omega, \alpha > \beta$
$=$	$\{\alpha, \alpha, \alpha\}$	$\alpha \in \Omega$

As $\mathcal{A}_{\mathbb{Q}_{\geq 0}}(\mathfrak{D})$ allows all finite-valued VCSPs, it is intractable even over a Boolean domain [67] as it includes the Max-SAT problem for the exclusive or predicate [87, 230].

The joint-winner property [80] for finite-valued VCSPs gives

Corollary 9.3 (of Theorem 9.1) $\mathcal{A}_{\mathbb{Q}_{\geq 0}}(\{<, =\})$ *is tractable.*

Cooper and Živný have shown that there are no other tractable cases; the results are depicted in Fig. 9.3: white nodes represent tractable cases and shaded nodes represent intractable cases.

Theorem 9.6 ([82]) *For* $|D| \geq 2$, *a class of binary finite-valued VCSP instances defined by* $\mathcal{A}_{\mathbb{Q}_{\geq 0}}(S)$, *where* $S \subseteq \{\triangle, <, >, =\}$, *is tractable if, and only if,* $S \subseteq \{<, =\}$.

Classification with Respect to Minimum Cost   The tractable classes $\mathcal{A}_{\{0,1\}}(\{>, 1\})$, $\mathcal{A}_{\{0,1\}}(\{>, 0\})$, and $\mathcal{A}_{\{0,1\}}(\{<, >\})$ appear in Fig. 9.2, but do not appear as subclasses of the tractable classes $\mathcal{A}_{\mathbb{Q}_{\geq 0}}(S)$ identified in Fig. 9.3. This is

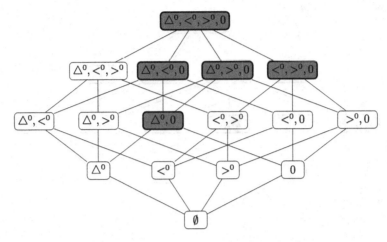

Fig. 9.4 Complexity of finite-valued VCSPs $\mathcal{A}_{\mathbb{Q}_{\geq 0}}(S)$, $S \subseteq \{\Delta^0, <^0, >^0, 0\}$

due to the fact that for the infinite set of possible costs $\Omega = \mathbb{Q}_{\geq 0}$, Fig. 9.3 covers only a subset of the infinite number of possible restrictions on triples of costs in triangles. We now consider triples of costs that allow us to find generalisations of these three tractable classes to finite-valued VCSPs, by considering restrictions depending on the relationship of costs with the minimum or maximum binary cost in an instance.

We start with the minimum cost. Without loss of generality we can assume that the minimum binary cost of an instance is 0. We consider the following types of triples of costs:

Symbol	Costs	Remark
Δ^0	$\{\alpha, \beta, 0\}$	$\alpha, \beta \in \Omega, \alpha > \beta > 0$
$<^0$	$\{0, 0, \alpha\}$	$\alpha \in \Omega, \alpha > 0$
$>^0$	$\{\alpha, \alpha, 0\}$	$\alpha \in \Omega, \alpha > 0$
0	$\{0, 0, 0\}$	

For simplicity of presentation, we do not consider the remaining type of triples of costs, namely $\{\alpha, \beta, \gamma\}$ such that $\alpha, \beta, \gamma > 0$. Since it is possible to transform any VCSP instance into an equivalent instance with nonzero costs by adding a constant $\epsilon > 0$ to all binary costs, it is clear that allowing all such triples of costs would render the VCSP intractable.

The complexity of combinations of costs from $\{\Delta^0, <^0, >^0, 0\}$ are shown in Fig. 9.4: white nodes represent tractable cases and shaded nodes represent intractable cases.

Theorem 9.7 ([82]) *For $|D| \geq 2$, a class of binary finite-valued VCSP instances defined by $\mathcal{A}_{\mathbb{Q}_{\geq 0}}(S)$, where $S \subseteq \{\Delta^0, <^0, >^0, 0\}$, is tractable if, and only if, $S \subseteq \{<^0, 0\}$, $S \subseteq \{>^0, 0\}$ or $S \subseteq \{\Delta^0, <^0, >^0\}$.*

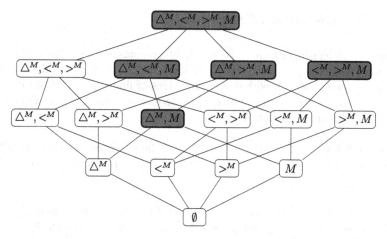

Fig. 9.5 Complexity of finite-valued VCSPs $\mathcal{A}_{\mathbb{Q}_{\geq 0}}(S)$, $S \subseteq \{\triangle^M, <^M, >^M, M\}$

Classification with Respect to Maximum Cost   Let $M \in \mathbb{Q}_{\geq 0}$ be any cost and consider the following types of triples of costs:

Symbol	Costs	Remark
$\triangle^M$	$\{\alpha, \beta, M\}$	$\alpha, \beta \in \Omega, \alpha < \beta < M$
$<^M$	$\{\alpha, \alpha, M\}$	$\alpha \in \Omega, \alpha < M$
$>^M$	$\{\alpha, M, M\}$	$\alpha \in \Omega, \alpha < M$
M	$\{M, M, M\}$	

Again, we do not consider the remaining type of triples of costs, namely $\{\alpha, \beta, \gamma\}$ such that $\alpha, \beta, \gamma < M$, since allowing such triples of costs renders the VCSP intractable. If $\{\triangle^M, <^M, >^M, M\}$ are the only allowed combinations of triples of costs, then M is clearly the maximum binary cost in the instance.

The complexity of combinations of costs from $\{\triangle^M, <^M, >^M, M\}$ are depicted in Fig. 9.5: white nodes represent tractable cases and shaded nodes represent intractable cases.

The most interesting case is $\mathcal{A}_{\mathbb{Q}_{\geq 0}}(\{>^M, M\})$, which turns out to be tractable by a reduction to maximum *weighted* matching [104] and hence is a proper generalisation of class $\mathcal{A}_{\{0,1\}}(\{>, 1\})$.

Theorem 9.8 ([82]) *For $|D| \geq 2$, a class of binary finite-valued VCSP instances defined by $\mathcal{A}_{\mathbb{Q}_{\geq 0}}(S)$, where $S \subseteq \{\triangle^M, <^M, >^M, M\}$, is tractable if, and only if, $S \subseteq \{<^M, M\}$ or $S \subseteq \{>^M, M\}$ or $S \subseteq \{\triangle^M, <^M, >^M\}$.*

9.2.4 General-Valued VCSPs

We now focus on general-valued VCSPs. In other words, we consider the complete valuation structure $\overline{\mathbb{Q}}_{\geq 0}$ as the set of possible costs Ω. In this case, the complexity classifications coincide with the classifications of finite-valued VCSPs obtained in Sect. 9.2.3: tractable cases remain tractable (see [82] for more details) and intractable cases remain intractable.

Theorem 9.9 ([82]) *For $|D| \geq 2$, a class of binary general-valued VCSP instances defined by $\mathcal{A}_{\overline{\mathbb{Q}}_{\geq 0}}(S)$, where $S \subseteq \{\Delta, <, >, =\}$, is tractable if, and only if, $S \subseteq \{<, =\}$.*

Theorem 9.10 ([82]) *For $|D| \geq 2$, a class of binary general-valued VCSP instances defined by $\mathcal{A}_{\overline{\mathbb{Q}}_{\geq 0}(S)}$, where $S \subseteq \{\Delta^0, <^0, >^0, 0\}$, is tractable if, and only if, $S \subseteq \{<^0, 0\}$, $S \subseteq \{>^0, 0\}$ or $S \subseteq \{\Delta^0, <^0, >^0\}$.*

Theorem 9.11 ([82]) *For $|D| \geq 2$, a class of binary general-valued VCSP instances defined by $\mathcal{A}_{\overline{\mathbb{Q}}_{\geq 0}}(S)$, where $S \subseteq \{\Delta^M, <^M, >^M, M\}$, is tractable if, and only if, $S \subseteq \{<^M, M\}$ or $S \subseteq \{>^M, M\}$ or $S \subseteq \{\Delta^M, <^M, >^M\}$.*

9.3 Cross-Free Convexity

In Sect. 9.2, we studied the computational complexity of several classes of binary VCSPs. In all considered cases, the joint-winner property (JWP) was either the only tractable case or one of only a few tractable cases.

In this section, we will generalise JWP to the *cross-free convexity property* (CFC) [82]. First we need to define some notation.

A function $g : \{0, \ldots, s\} \to \overline{\mathbb{Q}}_{\geq 0}$ is called *convex on the interval* $[l, u]$ if g is finite-valued on the interval $[l, u]$ and the derivative of g is nondecreasing on $[l, u]$, that is, $g(m+2) - g(m+1) \geq g(m+1) - g(m)$ for all $m = l, \ldots, u-2$. For brevity, we will say that g is *convex* if it is convex on some interval $[l, u] \subseteq [0, s]$ and infinite elsewhere (that is, on $[0, l-1] \cup [u+1, s]$).

Sets $A_1, \ldots, A_r \subseteq A$ are called *cross-free* if for every $1 \leq i, j \leq r$, either $A_i \subseteq A_j$, or $A_i \supseteq A_j$, or $A_i \cap A_j = \emptyset$, or $A_i \cup A_j = A$ [258].[3]

For notational convenience, we interpret a solution **x** (that is, an assignment to the variables $v_1, \ldots, v_n$) to a VCSP instance as the set of ⟨variable,value⟩ assignments $\{\langle v_i, x_i \rangle \mid x_i \in D_i \wedge i = 1, \ldots, n\}$.

If A_i is a set of ⟨variable,value⟩ assignments of a VCSP instance $\mathcal{P}$ and **x** a solution to $\mathcal{P}$, then we use the notation $|\mathbf{x} \cap A_i|$ to represent the number of ⟨variable, value⟩ assignments in the solution **x** that lie in A_i.

---

[3]Sets $A_1, \ldots, A_r$ satisfying only the first three conditions, that is, for every $1 \leq i, j \leq r$, $A_i \subseteq A_j$, or $A_i \supseteq A_j$, or $A_i \cap A_j = \emptyset$, are called *laminar* [258].

Definition 9.4 (Cross-free convexity) Let $\mathcal{P}$ be a VCSP instance. Let $A_1, \ldots, A_r$ be cross-free sets of ⟨variable,value⟩ assignments of $\mathcal{P}$. Let s_i be the number of distinct variables occurring in the set of ⟨variable, value⟩ assignments A_i. Instance $\mathcal{P}$ satisfies the *cross-free convexity property* if the objective function of $\mathcal{P}$ is $g(\mathbf{x}) = g_1(|\mathbf{x} \cap A_1|) + \cdots + g_r(|\mathbf{x} \cap A_r|)$, where each $g_i : [0, s_i] \to \mathbb{Q}$ $(i = 1, \ldots, r)$ is convex on an interval $[l_i, u_i] \subseteq [0, s_i]$ and $g_i(z) = \infty$ for $z \in [0, l_i - 1] \cup [u_i + 1, s_i]$.

Remark 9.3 Note that the functions g_i in Definition 9.4 are not the cost functions associated with the constraints.

Remark 9.4 The addition of any unary cost function cannot destroy the cross-free convexity property because for each ⟨variable,value⟩ assignment ⟨v_j, a⟩ we can add the singleton $A_i = \{\langle v_j, a \rangle\}$, which is necessarily either disjoint from or a subset of any other set A_k (and furthermore the corresponding function $g_i : \{0, 1\} \to \mathbb{Q}$ is trivially convex).

We now give a very special case of the cross-free convexity property, where all sets are disjoint and thus trivially cross-free.

Example 9.2 (Soft GCC) The GLOBAL CARDINALITY CONSTRAINT (GCC), introduced by Régin [243], is a generalisation of the ALLDIFFERENT constraint [242]. Given a set of n variables, the GCC specifies for each domain value d a lower bound l_d and an upper bound u_d on the number of variables that are assigned value d. The ALLDIFFERENT constraint is the special case of GCC with $l_d = 0$ and $u_d = 1$ for every d. Soft versions of the GCC have been considered by van Hoeve et al. [152].

The value-based soft GCC minimises the number of values below or above the given bound. We show that the value-based soft GCC satisfies the cross-free convexity property.

For every domain value $d \in D$, let $A_d = \{\langle v_i, d \rangle : i = 1, \ldots, n\}$. Clearly, $A_1, \ldots, A_s$ are disjoint, where $s = |D|$. For every d, let

$$g_d(m) \stackrel{\text{def}}{=} \begin{cases} l_d - m & \text{if } m < l_d, \\ 0 & \text{if } l_d \leq m \leq u_d, \\ m - u_d & \text{if } m > u_d. \end{cases}$$

It follows readily from the definition of g_d that the sequence $g_d(m + 1) - g_d(m)$, for $m = 0, \ldots, n - 1$, which is the derivation of g_d, is the sequence $-1, \ldots, -1, 0, \ldots, 0, 1, \ldots, 1$. Therefore, for every d, g_d has a nondecreasing derivative and hence is convex.

Cooper and Živný have shown, using a reduction to the minimum convex cost flow problem [2], that the CFC defines a tractable class of VCSPs.

Theorem 9.12 ([82]) *Any VCSP instance $\mathcal{P}$ satisfying the cross-free convexity property can be solved in polynomial time.*

We note that relaxing either convexity or cross-freeness in Definition 9.4 leads to intractability.

Theorem 9.13 ([82]) *The class of VCSP instances whose objective function is of the form $g(\mathbf{x}) = g_1(|\mathbf{x} \cap A_1|) + \cdots + g_r(|\mathbf{x} \cap A_r|)$, where the functions g_i are convex but the sets of assignments A_i are not necessarily cross-free, is NP-hard, even if $|A_i| \leq 2$ for all $i \in \{1, \ldots, r\}$ and all variables are Boolean.*

Theorem 9.14 ([82]) *The class of VCSP instances whose objective function is of the form $g(\mathbf{x}) = g_1(|\mathbf{x} \cap A_1|) + \cdots + g_r(|\mathbf{x} \cap A_r|)$, where the sets of assignments A_i are cross-free but the functions g_i are not necessarily convex, is NP-hard even if $|A_i| \leq 3$ for all $i \in \{1, \ldots, r\}$ and all variables are Boolean.*

(In fact, the above two theorems hold in a slightly stronger setting; see [82] for more details.)

Remark 9.5 Note that the NP-hardness result in Theorem 9.14 requires assignment sets of size up to 3. It turns out that the class of VCSP instances with cross-free assignment sets of size at most 2 and arbitrary functions is tractable. Cooper and Živný have shown that (i) this class is tractable (of bounded treewidth) with Boolean domains, and that (ii) problems with domains of size $d > 3$ are polynomial-time equivalent to problems with domains of size 3 [82]. For domains of size 3, this class has recently been shown tractable in [77].

9.4   Summary

In this chapter we have studied hybrid tractability of VCSPs. In particular, in Sect. 9.2, we have surveyed recent work on the tractability of classes of instances defined by properties of subproblems of size $k = 3$.[4]

9.4.1   Related Work

Hybrid tractability has been studied in the context of CSPs. Cohen observed that solving a CSP instance $\mathcal{P}$ is equivalent to finding a maximum independent set in the *microstructure* complement [169] of $\mathcal{P}$. Since the latter is known to be tractable for triangulated (also known as chordal) graphs [246], and triangulated graphs are recognisable in polynomial time [263], CSP instances with triangulated microstructure complements are tractable. Trivially, the same result holds for any other class

---

[4]For $k = 2$, such properties can only define language classes.

of graphs that is polynomial-time recognisable and admits a polynomial-time algorithm for the maximum independent set problem. More results on hybrid tractability of CSPs can be found in [64, 74, 76, 115, 193, 220, 273].

Up until now there have been very few results on hybrid tractability for VCSPs. In fact, [80, 82] are the only works we are aware of. For instance, Kumar defines an interesting framework for hybrid tractability for the Boolean weighted CSP [194]. However, to the best of our knowledge, this framework has so far not provided any new hybrid classes. In fact, all tractable classes presented in [194] are not hybrid and were already known.

9.4.2 Open Problems

An interesting and widely open question is that of what other hybrid classes can be defined by properties of k-variable subproblems for $k \geq 3$. We have presented one such hybrid class. Generalising the joint-winner property (JWP), which is defined by a subproblem of size 3 and which plays an important role in the complexity classification of VCSPs defined by subproblems of size 3, we have described in Sect. 9.3 a large class of tractable VCSPs of unbounded arity. What other hybrid classes of VCSPs are there?

Chapter 10
Summary and Open Problems

*In mathematics the art of proposing a question must be held of
higher value than solving it.*
Georg Cantor

10.1 Summary

This book has studied the minimisation problem given by a sum of fixed-arity functions over finite discrete domains. As we have seen, such problems can be cast as
valued constraint satisfaction problems and also as optimisation problems in several
other well-studied frameworks.

In the second part of the book, we have given a detailed examination of the expressive power of cost functions (valued constraints). First, we have presented a
general algebraic theory for the expressive power of valued constraints and established a Galois connection. Second, we have studied several classes of cost functions
and their expressive power.

In the third part of the book, we have focused on the tractability of the minimisation problem of functions given by a sum of fixed-arity functions. First, we have
presented a list of all known tractable languages. Second, we have given a summary
of recent results on conservative languages. Next, we have studied the power of
linear programming. Finally, we have surveyed recent results on hybrid tractability.

A concept that has turned out important at several places is the concept of submodularity, often regarded as the discrete analogue of convexity, and its various
generalisations.

10.2 Open Problems

The end of each chapter mentions related work and open problems. We finish with
some more general open problems.

1. What is the structure of weighted clones?
2. Which valued constraint languages are tractable?
3. Which fractional polymorphisms guarantee tractability?

S. Živný, *The Complexity of Valued Constraint Satisfaction Problems*,
Cognitive Technologies, DOI 10.1007/978-3-642-33974-5_10,
© Springer-Verlag Berlin Heidelberg 2012

4. Are all tractable languages solvable by the basic LP relaxation?
5. What are the extreme rays of (k-ary) submodular functions?
6. Which functions are expressible by binary submodular functions?
7. Which classes of submodular functions are closed under expressibility?
8. Which classes of VCSPs are captured by subproblems on k-variables?
9. What other hybrid classes of VCSPs are there?
10. What other optimisation problems are tractable (with respect to exact solvability and other notions of tractability) for functions captured by fractional polymorphisms (for both explicitly given functions and functions given by oracles)?

References

If I have seen further it is by standing on the shoulders of giants.
Isaac Newton

1. Adler, I., Gottlob, G., Grohe, M.: Hypertree width and related hypergraph invariants. Eur. J. Comb. **28**(8), 2167–2181 (2007)
2. Ahuja, R., Magnanti, T., Orlin, J.: Network Flows: Theory, Algorithms, and Applications. Prentice Hall/Pearson, Upper Saddle River (2005)
3. Apt, K.: Principles of Constraint Programming. Cambridge University Press, Cambridge (2003)
4. Atserias, A., Bulatov, A.A., Dalmau, V.: On the power of k-consistency. In: Proceedings of the 34th International Colloquium on Automata, Languages and Programming (ICALP'07). Lecture Notes in Computer Science, vol. 4596, pp. 279–290. Springer, Berlin (2007)
5. Atserias, A., Weyer, M.: Decidable relationships between consistency notions for constraint satisfaction problems. In: Proceedings of the 18th Annual Conference of the European Association for Computer Science Logic (CSL'09). Lecture Notes in Computer Science, vol. 5771, pp. 102–116. Springer, Berlin (2009)
6. Ausiello, G., Crescenzi, P., Gambosi, G., Kann, V., Marchetti-Spaccamela, A., Protasi, M.: Complexity and Approximation: Combinatorial Optimization Problems and Their Approximability Properties. Springer, Berlin (1999)
7. Avis, D., Kaluzny, B.: Solving inequalities and proving Farkas's lemma made easy. Am. Math. Mon. **111**(2), 152–157 (2004)
8. Balinksi, M.: On a selection problem. Manag. Sci. **17**(3), 230–231 (1970)
9. Bang-Jensen, J., Hell, P., MacGillivray, G.: The complexity of colouring by semicomplete digraphs. SIAM J. Discrete Math. **1**(3), 281–298 (1988)
10. Barto, L.: The dichotomy for conservative constraint satisfaction problems revisited. In: Proceedings of the 26th IEEE Symposium on Logic in Computer Science (LICS'11), pp. 301–310. IEEE Computer Society, Los Alamitos (2011)
11. Barto, L., Kozik, M.: Constraint satisfaction problems of bounded width. In: Proceedings of the 50th Annual IEEE Symposium on Foundations of Computer Science (FOCS'09), pp. 461–471. IEEE Computer Society, Los Alamitos (2009)
12. Barto, L., Kozik, M.: Robust satisfiability of constraint satisfaction problems. In: Proceedings of the 44th Annual ACM Symposium on Theory of Computing (STOC'12), pp. 931–940. ACM, New York (2012)
13. Barto, L., Kozik, M., Maróti, M., Niven, T.: CSP dichotomy for special triads. Proc. Am. Math. Soc. **137**(9), 2921–2934 (2009)
14. Barto, L., Kozik, M., Niven, T.: The CSP dichotomy holds for digraphs with no sources and no sinks (a positive answer to a conjecture of Bang-Jensen and Hell). SIAM J. Sci. Comput. **38**(5), 1782–1802 (2009)
15. van Beek, P., Dechter, R.: On the minimality and decomposability of row-convex constraint networks. J. ACM **42**(3), 543–561 (1995)

16. Berman, J., Idziak, P., Marković, P., McKenzie, R., Valeriote, M., Willard, R.: Varieties with few subalgebras of powers. Trans. Am. Math. Soc. **362**(3), 1445–1473 (2010)

17. Bertelé, U., Brioshi, F.: Nonserial Dynamic Programming. Academic Press, San Diego (1972)

18. Besag, J.: On the statistical analysis of dirty pictures. J. R. Stat. Soc. B **48**(3), 259–302 (1986)

19. Bilbao, J.M., Fernández, J.R., Jiménez, N., López, J.J.: Survey of bicooperative games. In: Chinchuluun, A., Pardalos, P.M., Migdalas, A., Pitsoulis, L. (eds.) Pareto Optimality, Game Theory and Equilibria. Springer, Berlin (2008)

20. Billionet, A., Minoux, M.: Maximizing a supermodular pseudo-Boolean function: a polynomial algorithm for cubic functions. Discrete Appl. Math. **12**(1), 1–11 (1985)

21. Bistarelli, S., Montanari, U., Rossi, F., Schiex, T., Verfaillie, G., Fargier, H.: Semiring-based CSPs and valued CSPs: frameworks, properties, and comparison. Constraints **4**(3), 199–240 (1999)

22. Bodirsky, M.: Constraint satisfaction problems with infinite templates. In: Complexity of Constraints. Lecture Notes in Computer Science, vol. 5250, pp. 196–228. Springer, Berlin (2008)

23. Bodirsky, M., Chen, H.: Quantified equality constraints. In: Proceedings of the 22nd IEEE Symposium on Logic in Computer Science (LICS'07), pp. 203–212 (2007)

24. Bodirsky, M., Chen, H.: Relatively quantified constraint satisfaction. Constraints **14**(1), 3–15 (2009)

25. Bodirsky, M., Kára, J.: The complexity of equality constraint languages. Theory Comput. Syst. **43**(2), 136–158 (2008)

26. Bodirsky, M., Kára, J.: The complexity of temporal constraint satisfaction problems. Journal of the ACM **57**(2) (2010)

27. Bodnarčuk, V., Kalužnin, L., Kotov, V., Romov, B.: Galois theory for Post algebras. I. Cybern. Syst. Anal. **5**(3), 243–252 (1969)

28. Böhler, E., Creignou, N., Reith, S., Vollmer, H.: Playing with Boolean blocks, part I: Post's lattice with applications to complexity theory. ACM SIGACT Newslett. **34**(4), 38–52 (2003)

29. Böhler, E., Creignou, N., Reith, S., Vollmer, H.: Playing with Boolean blocks, part II: Constraint satisfaction problems. ACM SIGACT Newslett. **35**(1), 22–35 (2004)

30. Böhler, E., Reith, S., Schnoor, H., Vollmer, H.: Bases for Boolean co-clones. Inf. Process. Lett. **96**(2), 59–66 (2005)

31. Börner, F.: Basics of Galois connections. In: Complexity of Constraints. Lecture Notes in Computer Science, vol. 5250, pp. 38–67. Springer, Berlin (2008)

32. Börner, F., Bulatov, A., Jeavons, P., Krokhin, A.: Quantified constraints: algorithms and complexity. In: Proceedings of Computer Science Logic, the 17th Inernational Workshop (CSL'03), the 12th Annual Conference of the EACSL, and the 8th Kurt Gödel Colloquium. Lecture Notes in Computer Science, vol. 2803, pp. 58–70. Springer, Berlin (2003)

33. Boros, E., Gruber, A.: On quadratization of pseudo-Boolean functions. In: Proceedings of the International Symposium on Artificial Intelligence and Mathematics (ISAIM'12) (2012)

34. Boros, E., Hammer, P.L.: Pseudo-Boolean optimization. Discrete Appl. Math. **123**(1–3), 155–225 (2002)

35. Bouchet, A.: Greedy algorithm and symmetric matroids. Math. Program. **38**(1), 147–159 (1987)

36. Boykov, Y., Huttenlocher, D.P.: A new Bayesian framework for object recognition. In: IEEE Computer Society Conference on Computer Vision and Pattern Recognition (CVPR'99), pp. 2517–2523. IEEE Computer Society, Los Alamitos (1999)

37. Boykov, Y., Jolly, M.P.: Interactive graph cuts for optimal boundary and region segmentation of objects in N-D images. In: Proceedings of the Eighth International Conference on Computer Vision (ICCV'01), pp. 105–112 (2001)

38. Bulatov, A.: A graph of a relational structure and constraint satisfaction problems. In: Proceedings 19th IEEE Symposium on Logic in Computer Science (LICS'04), pp. 448–457. IEEE Computer Society, Los Alamitos (2004)

39. Bulatov, A.: A dichotomy theorem for constraint satisfaction problems on a 3-element set. J. ACM **53**(1), 66–120 (2006)
40. Bulatov, A., Dalmau, V.: A simple algorithm for Mal'tsev constraints. SIAM J. Sci. Comput. **36**(1), 16–27 (2006)
41. Bulatov, A., Krokhin, A., Jeavons, P.: The complexity of maximal constraint languages. In: Proceedings 33rd ACM Symposium on Theory of Computing (STOC'01), pp. 667–674 (2001)
42. Bulatov, A., Krokhin, A., Jeavons, P.: Classifying the complexity of constraints using finite algebras. SIAM J. Sci. Comput. **34**(3), 720–742 (2005)
43. Bulatov, A.A.: H-Coloring dichotomy revisited. Theor. Comput. Sci. **349**(1), 31–39 (2005)
44. Bulatov, A.A.: The complexity of the counting constraint satisfaction problem. In: Proceedings of the 35th International Colloquium on Automata, Languages and Programming (ICALP'08). Lecture Notes in Computer Science, vol. 5126, pp. 646–661. Springer, Berlin (2008)
45. Bulatov, A.A.: Complexity of conservative constraint satisfaction problems. ACM Trans. Comput. Log. **12**(4), 24 (2011)
46. Bulatov, A.A.: On the CSP dichotomy conjecture. In: Proceedings of the 6th International Computer Science Symposium in Russia (CSR'11). Lecture Notes in Computer Science, vol. 6651, pp. 331–344. Springer, Berlin (2011). Invited paper
47. Bulatov, A.A., Dyer, M.E., Goldberg, L.A., Jerrum, M.: Log-supermodular functions, functional clones and counting CSPs. In: Proceedings of the 29th International Symposium on Theoretical Aspects of Computer Science (STACS'12), pp. 302–313 (2012)
48. Bulatov, A.A., Krokhin, A., Larose, B.: Dualities for constraint satisfaction problems. In: Complexity of Constraints. Lecture Notes in Computer Science, vol. 5250, pp. 93–124. Springer, Berlin (2008)
49. Burkard, R., Klinz, B., Rudolf, R.: Perspectives of Monge properties in optimization. Discrete Appl. Math. **70**(2), 95–161 (1996)
50. Cai, J.Y., Chen, X.: Complexity of counting CSP with complex weights. In: Proceedings of the 44th Annual ACM Symposium on Theory of Computing (STOC'12), pp. 909–920. ACM, New York (2012)
51. Chandrasekaran, R., Kabadi, S.N.: Pseudomatroids. Discrete Math. **71**(3), 205–217 (1988)
52. Charpiat, G.: Exhaustive family of energies minimizable exactly by a graph cut. In: Proceedings of the 24th IEEE Conference on Computer Vision and Pattern Recognition (CVPR'11), pp. 1849–1856. IEEE, New York (2011)
53. Chen, H.: The computational complexity of quantified constraint satisfaction. Ph.D. thesis, Cornell University (2004)
54. Chen, H.: A rendezvous of logic, complexity, and algebra. SIGACT News **37**(4), 85–114 (2006)
55. Chen, H.: The complexity of quantified constraint satisfaction: collapsibility, sink algebras, and the three-element case. SIAM J. Comput. **37**(5), 1674–1701 (2008)
56. Chen, H., Dalmau, V.: Beyond hypertree width: decomposition methods without decompositions. In: Proceedings of the 11th International Conference on Principles and Practice of Constraint Programming (CP'05). Lecture Notes in Computer Science, vol. 3709, pp. 167–181. Springer, Berlin (2005)
57. Chen, X., Dyer, M., Goldberg, L.A., Jerrum, M., Lu, P., McQuillan, C., Richerby, D.: The complexity of approximating conservative counting CSPs. Tech. rep. (2012). arXiv: 1208.1783
58. Cohen, D., Cooper, M., Jeavons, P., Krokhin, A.: A maximal tractable class of soft constraints. J. Artif. Intell. Res. **22**, 1–22 (2004)
59. Cohen, D., Cooper, M., Jeavons, P., Krokhin, A.: Supermodular functions and the complexity of MAX-CSP. Discrete Appl. Math. **149**(1–3), 53–72 (2005)
60. Cohen, D., Jeavons, P.: The complexity of constraint languages. In: Rossi, F., van Beek, P., Walsh, T. (eds.) The Handbook of Constraint Programming. Elsevier, Amsterdam (2006)

61. Cohen, D., Jeavons, P., Gyssens, M.: A unified theory of structural tractability for constraint satisfaction problems. J. Comput. Syst. Sci. **74**(5), 721–743 (2008)

62. Cohen, D.A.: A new class of binary CSPs for which arc-consistency is a decision procedure. In: Proceedings of the 9th International Conference on Principles and Practice of Constraint Programming (CP'03). Lecture Notes in Computer Science, vol. 2833, pp. 807–811. Springer, Berlin (2003)

63. Cohen, D.A., Cooper, M.C., Creed, P., Jeavons, P., Živný, S.: An algebraic theory of complexity for discrete optimisation. Tech. rep. (2012). arXiv:1207.6692

64. Cohen, D.A., Cooper, M.C., Green, M., Marx, D.: On guaranteeing polynomially-bounded search tree size. In: Proceedings of the 17th International Conference on Principles and Practice of Constraint Programming (CP'11). Lecture Notes in Computer Science, vol. 6876, pp. 160–171. Springer, Berlin (2011)

65. Cohen, D.A., Cooper, M.C., Jeavons, P.G.: An algebraic characterisation of complexity for valued constraints. In: Proceedings of the 12th International Conference on Principles and Practice of Constraint Programming (CP'06). Lecture Notes in Computer Science, vol. 4204, pp. 107–121. Springer, Berlin (2006)

66. Cohen, D.A., Cooper, M.C., Jeavons, P.G.: Generalising submodularity and Horn clauses: tractable optimization problems defined by tournament pair multimorphisms. Theor. Comput. Sci. **401**(1–3), 36–51 (2008)

67. Cohen, D.A., Cooper, M.C., Jeavons, P.G., Krokhin, A.A.: The complexity of soft constraint satisfaction. Artif. Intell. **170**(11), 983–1016 (2006)

68. Cohen, D.A., Creed, P., Jeavons, P.G., Živný, S.: An algebraic theory of complexity for valued constraints: establishing a Galois connection. In: Proceedings of the 36th International Symposium on Mathematical Foundations of Computer Science (MFCS'11). Lecture Notes in Computer Science, vol. 6907, pp. 231–242. Springer, Berlin (2011)

69. Cohen, D.A., Jeavons, P.G., Živný, S.: The expressive power of valued constraints: hierarchies and collapses. In: Proceedings of the 13th International Conference on Principles and Practice of Constraint Programming (CP'07). Lecture Notes in Computer Science, vol. 4741, pp. 798–805. Springer, Berlin (2007)

70. Cohen, D.A., Jeavons, P.G., Živný, S.: The expressive power of valued constraints: hierarchies and collapses. Theor. Comput. Sci. **409**(1), 137–153 (2008)

71. Cooper, M.C.: High-order consistency in valued constraint satisfaction. Constraints **10**(3), 283–305 (2005)

72. Cooper, M.C.: Line Drawing Interpretation. Springer, Berlin (2008)

73. Cooper, M.C.: Minimization of locally defined submodular functions by optimal soft arc consistency. Constraints **13**(4), 437–458 (2008)

74. Cooper, M.C., Escamocher, G.: A dichotomy for 2-constraint forbidden CSP patterns. In: Proceedings of AAAI'12. AAAI Press, Menlo Park (2012)

75. Cooper, M.C., de Givry, S., Sánchez, M., Schiex, T., Zytnicki, M., Werner, T.: Soft arc consistency revisited. Artif. Intell. **174**(7–8), 449–478 (2010)

76. Cooper, M.C., Jeavons, P.G., Salamon, A.Z.: Generalizing constraint satisfaction on trees: hybrid tractability and variable elimination. Artif. Intell. **174**(9–10), 570–584 (2010)

77. Cooper, M.C., Živný, S.: Tractability of cross-free and non-convex VCSPs. Unpublished Manuscript (2012)

78. Cooper, M.C., Živný, S.: A new hybrid tractable class of soft constraint problems. In: Proceedings of the 16th International Conference on Principles and Practice of Constraint Programming (CP'10). Lecture Notes in Computer Science, vol. 6308, pp. 152–166. Springer, Berlin (2010)

79. Cooper, M.C., Živný, S.: Hierarchically nested convex VCSP. In: Proceedings of the 17th International Conference on Principles and Practice of Constraint Programming (CP'11). Lecture Notes in Computer Science, vol. 6876, pp. 187–194. Springer, Berlin (2011)

80. Cooper, M.C., Živný, S.: Hybrid tractability of valued constraint problems. Artif. Intell. **175**(9–10), 1555–1569 (2011)

81. Cooper, M.C., Živný, S.: Tractable triangles. In: Proceedings of the 17th International Conference on Principles and Practice of Constraint Programming (CP'11). Lecture Notes in Computer Science, vol. 6876, pp. 195–209. Springer, Berlin (2011)

82. Cooper, M.C., Živný, S.: Tractable triangles and cross-free convexity in discrete optimisation. J. Artif. Intell. Res. **44**, 455–490 (2012)

83. Crama, Y.: Recognition problems for special classes of polynomials in 0–1 variables. Math. Program. **44**(1–3), 139–155 (1989)

84. Crama, Y., Hammer, P.L.: Boolean Functions—Theory, Algorithms, and Applications. Cambridge University Press, Cambridge (2011)

85. Creed, P., Živný, S.: On minimal weighted clones. In: Proceedings of the 17th International Conference on Principles and Practice of Constraint Programming (CP'11). Lecture Notes in Computer Science, vol. 6876, pp. 210–224. Springer, Berlin (2011)

86. Creed, P., Živný, S.: Conservative valued CSPs revisited. Unpublished Manuscript (2012)

87. Creignou, N., Khanna, S., Sudan, M.: Complexity Classification of Boolean Constraint Satisfaction Problems. SIAM Monographs on Discrete Mathematics and Applications, vol. 7. SIAM, Philadelphia (2001)

88. Creignou, N., Kolaitis, P.G., Vollmer, H. (eds.): Complexity of Constraints: An Overview of Current Research Themes. Lecture Notes in Computer Science, vol. 5250. Springer, Berlin (2008)

89. Creignou, N., Kolaitis, P.G., Zanuttini, B.: Structure identification of Boolean relations and plain bases for co-clones. J. Comput. Syst. Sci. **74**(7), 1103–1115 (2008)

90. Cunningham, W.H.: Testing membership in matroid polyhedra. J. Comb. Theory, Ser. B **36**(2), 161–188 (1984)

91. Cunningham, W.H.: On submodular function minimization. Combinatorica **5**(3), 185–192 (1985)

92. Dalmau, V.: Generalized majority-minority operations are tractable. Log. Methods Comput. Sci. **2**(4) (2006)

93. Dalmau, V., Kolaitis, P.G., Vardi, M.Y.: Constraint satisfaction, bounded treewidth, and finite-variable logics. In: Proceedings of the 8th International Conference on Principles and Practice of Constraint Programming (CP'02). Lecture Notes in Computer Science, vol. 2470, pp. 310–326. Springer, Berlin (2002)

94. Dalmau, V., Pearson, J.: Set functions and width 1 problems. In: Proceedings of the 5th International Conference on Constraint Programming (CP'99). Lecture Notes in Computer Science, vol. 1713, pp. 159–173. Springer, Berlin (1999)

95. Dechter, R.: On the expressiveness of networks with hidden variables. In: Proceedings of the 8th National Conference on Artificial Intelligence (AAAI'90), pp. 556–562 (1990)

96. Dechter, R.: From local to global consistency. Artif. Intell. **55**(1), 87–107 (1992)

97. Dechter, R.: Constraint Processing. Morgan Kaufmann, San Mateo (2003)

98. Dechter, R., Pearl, J.: Tree clustering for constraint networks. Artif. Intell. **38**(3), 353–366 (1989)

99. Dechter, R., Pearl, J.: Structure identification in relational data. Artif. Intell. **58**(1–3), 237–270 (1992)

100. Deineko, V., Jonsson, P., Klasson, M., Krokhin, A.: The approximability of Max CSP with fixed-value constraints. J. ACM **55**(4) (2008)

101. Denecke, K., Wismath, S.L.: Universal Algebra and Applications in Theoretical Computer Science. Chapman and Hall/CRC Press, London/Boca Raton (2002)

102. Downey, R., Fellows, M.: Parametrized Complexity. Springer, Berlin (1999)

103. Dyer, M.E., Richerby, D.: On the complexity of #CSP. In: Proceedings of the 42nd ACM Symposium on Theory of Computing (STOC'10), pp. 725–734. ACM, New York (2010)

104. Edmonds, J.: Maximum matching and a polyhedron with 0, 1 vertices. J. Res. Natl. Bur. Stand. B **69**, 125–130 (1965)

105. Edmonds, J.: Paths, trees, and flowers. Can. J. Math. **17**, 449–467 (1965)

106. Edmonds, J.: Submodular functions, matroids, and certain polyhedra. In: Combinatorial Structures and Their Applications, pp. 69–87 (1970)

107. Färnqvist, T., Jonsson, P.: Bounded tree-width and CSP-related problems. In: Proceedings of the 18th International Symposium on Algorithms and Computation (ISAAC'07). Lecture Notes in Computer Science, vol. 4835, pp. 632–643. Springer, Berlin (2007)

108. Fearnley, A.: A strongly rigid binary relation. Acta Sci. Math. (Szeged) **61**(1–4), 35–41 (1995)

109. Feder, T., Hell, P., Huang, J.: Bi-arc graphs and the complexity of list homomorphisms. J. Graph Theory **42**(1), 61–80 (2003)

110. Feder, T., Kolaitis, P.: Closures and dichotomies for quantified constraints. Tech. rep. TR06-160, Electronic Colloquium on Computational Complexity (ECCC) (2006)

111. Feder, T., Vardi, M.Y.: The computational structure of monotone monadic SNP and constraint satisfaction: a study through datalog and group theory. SIAM J. Sci. Comput. **28**(1), 57–104 (1998)

112. Feige, U.: A threshold of ln n for approximating set cover. J. ACM **45**(4), 634–652 (1998)

113. Feige, U., Mirrokni, V.S., Vondrák, J.: Maximizing non-monotone submodular functions. In: Proceedings of the 48th Annual IEEE Symposium on Foundations of Computer Science (FOCS'07), pp. 461–471. IEEE Computer Society, Los Alamitos (2007)

114. Feige, U., Mirrokni, V.S., Vondrák, J.: Maximizing non-monotone submodular functions. SIAM J. Comput. **40**(4), 1133–1153 (2011)

115. Fellows, M.R., Friedrich, T., Hermelin, D., Narodytska, N., Rosamond, F.A.: Constraint satisfaction problems: convexity makes all different constraints tractable. In: Proceedings of the 22nd International Joint Conference on Artificial Intelligence (IJCAI'11), pp. 522–527 (2011)

116. Fleischer, L., Iwata, S.: A push-relabel framework for submodular function minimization and applications to parametric optimization. Discrete Appl. Math. **131**(2), 311–322 (2003)

117. Flum, J., Grohe, M.: Parametrized Complexity Theory. Texts in Theoretical Computer Science. An EATCS Series. Springer, Berlin (2006)

118. Freuder, E.C.: Synthesizing constraint expressions. Commun. ACM **21**(11), 958–966 (1978)

119. Freuder, E.C.: A sufficient condition for backtrack-bounded search. J. ACM **32**, 755–761 (1985)

120. Freuder, E.C.: Complexity of K-tree structured constraint satisfaction problems. In: Proceedings of the 8th National Conference on Artificial Intelligence (AAAI'90), pp. 4–9 (1990)

121. Fujishige, S.: Submodular Functions and Optimization, 2nd edn. Annals of Discrete Mathematics, vol. 58. North-Holland, Amsterdam (2005)

122. Fujishige, S., Iwata, S.: Bisubmodular function minimization. SIAM J. Discrete Math. **19**(4), 1065–1073 (2005)

123. Fujishige, S., Patkar, S.B.: Realization of set functions as cut functions of graphs and hypergraphs. Discrete Math. **226**(1–3), 199–210 (2001)

124. Gallo, G., Simeone, B.: On the supermodular knapsack problem. Math. Program. **45**(1–3), 295–309 (1988)

125. Garey, M.R., Johnson, D.S.: Computers and Intractability: A Guide to the Theory of NP-Completeness. W.H. Freeman, New York (1979)

126. Gault, R., Jeavons, P.: Implementing a test for tractability. Constraints **9**(2), 139–160 (2004)

127. Geiger, D.: Closed systems of functions and predicates. Pac. J. Math. **27**(1), 95–100 (1968)

128. Geman, S., Geman, D.: Stochastic relaxation, Gibbs distributions, and the Bayesian restoration of images. IEEE Trans. Pattern Anal. Mach. Intell. **6**(6), 721–741 (1984)

129. Gil, A., Hermann, M., Salzer, G., Zanuttini, B.: Efficient algorithms for description problems over finite totally ordered domains. SIAM J. Comput. **38**(3), 922–945 (2008)

130. Goemans, M.X., Williamson, D.P.: Improved approximation algorithms for maximum cut and satisfiability problems using semidefinite programming. J. ACM **42**(6), 1115–1145 (1995)

131. Goldberg, A.V., Tarjan, R.E.: A new approach to the maximum flow problem. J. ACM **35**(4), 921–940 (1988)

132. Gottlob, G., Greco, G., Scarcello, F.: Tractable optimization problems through hypergraph-based structural restrictions. In: Proceedings of the 36th International Colloquium on Au-

tomata, Languages and Programming (ICALP'09, Part II). Lecture Notes in Computer Science, vol. 5556, pp. 16–30. Springer, Berlin (2007)

133. Gottlob, G., Leone, N., Scarcello, F.: A comparison of structural CSP decomposition methods. Artif. Intell. **124**(2), 243–282 (2000)

134. Gottlob, G., Leone, N., Scarcello, F.: Hypertree decomposition and tractable queries. J. Comput. Syst. Sci. **64**(3), 579–627 (2002)

135. Gottlob, G., Miklós, Z., Schwentick, T.: Generalized hypertree decompositions: NP-hardness and tractable variants. J. ACM **56**(6) (2009)

136. Gottlob, G., Szeider, S.: Fixed-parameter algorithms for artificial intelligence, constraint satisfaction and database problems. Comput. J. **51**(3), 303–325 (2008)

137. Grädel, E., Kolaitis, P.G., Libkin, L., Marx, M., Spencer, J., Vardi, M.Y., Venema, Y., Weinstein, S.: Finite Model Theory and Its Applications. Texts in Theoretical Computer Science. An EATCS Series. Springer, Berlin (2007)

138. Green, M.J., Cohen, D.A.: Domain permutation reduction for constraint satisfaction problems. Artif. Intell. **172**(8–9), 1094–1118 (2008)

139. Grohe, M.: The complexity of homomorphism and constraint satisfaction problems seen from the other side. J. ACM **54**(1), 1–24 (2007)

140. Grohe, M., Marx, D.: Constraint solving via fractional edge covers. In: Proceedings of the 17th Annual ACM-SIAM Symposium on Discrete Algorithms (SODA'06), pp. 289–298. SIAM, Philadelphia (2006)

141. Grötschel, M., Lovasz, L., Schrijver, A.: The ellipsoid method and its consequences in combinatorial optimization. Combinatorica **1**(2), 169–198 (1981)

142. Grötschel, M., Lovasz, L., Schrijver, A.: Geometric Algorithms and Combinatorial Optimization. Algorithms and Combinatorics, vol. 2. Springer, Berlin (1988)

143. Gutin, G., Hell, P., Rafiey, A., Yeo, A.: A dichotomy for minimum cost graph homomorphisms. Eur. J. Comb. **29**(4), 900–911 (2008)

144. Gutin, G., Kim, E.: Introduction to the minimum cost homomorphism problem for directed and undirected graphs. Lect. Notes Ramanujan Math. Soc. **7**, 25–37 (2008)

145. Gutin, G., Rafiey, A., Yeo, A., Tso, M.: Level of repair analysis and minimum cost homomorphisms of graphs. Discrete Appl. Math. **154**(6), 881–889 (2006)

146. Gyssens, M., Jeavons, P.G., Cohen, D.A.: Decomposing constraint satisfaction problems using database techniques. Artif. Intell. **66**(1), 57–89 (1994)

147. Hammer, P.L.: Some network flow problems solved with pseudo-Boolean programming. Oper. Res. **13**(3), 388–399 (1965)

148. Hansen, P., Simeone, B.: Unimodular functions. Discrete Appl. Math. **14**(3), 269–281 (1986)

149. Hell, P., Nešetřil, J.: On the complexity of H-coloring. J. Comb. Theory, Ser. B **48**(1), 92–110 (1990)

150. Hell, P., Nešetřil, J.: Graphs and Homomorphisms. Oxford University Press, London (2004)

151. Hell, P., Nešetřil, J.: Colouring, constraint satisfaction, and complexity. Comput. Sci. Rev. **2**(3), 143–163 (2008)

152. van Hoeve, W.J., Pesant, G., Rousseau, L.M.: On global warming: flow-based soft global constraints. J. Heuristics **12**(4–5), 347–373 (2006)

153. Huber, A., Kolmogorov, V.: Towards minimizing k-submodular functions. In: Proceedings of the 2nd International Symposium on Combinatorial Optimization (ISCO'12) (2012)

154. Huber, A., Krokhin, A., Powell, R.: Skew bisubmodularity and valued CSPs. In: Proceedings of the 24th ACM-SIAM Symposium on Discrete Algorithms (SODA'13). SIAM, Philadelphia (2012)

155. Idziak, P.M., Markovic, P., McKenzie, R., Valeriote, M., Willard, R.: Tractability and learnability arising from algebras with few subpowers. SIAM J. Comput. **39**(7), 3023–3037 (2010)

156. Iwata, S.: A fully combinatorial algorithm for submodular function minimization. J. Comb. Theory, Ser. B **84**(2), 203–212 (2002)

157. Iwata, S.: A faster scaling algorithm for minimizing submodular functions. SIAM J. Sci. Comput. **32**(4), 833–840 (2003)

158. Iwata, S.: Submodular function minimization. Math. Program. **112**(1), 45–64 (2008)
159. Iwata, S., Fleischer, L., Fujishige, S.: A combinatorial strongly polynomial algorithm for minimizing submodular functions. J. ACM **48**(4), 761–777 (2001)
160. Iwata, S., Orlin, J.B.: A simple combinatorial algorithm for submodular function minimization. In: Proceedings of the 20th Annual ACM-SIAM Symposium on Discrete Algorithms (SODA'09), pp. 1230–1237. SIAM, Philadelphia (2009)
161. Jeavons, P.: Presenting constraints. In: Proceedings of the 18th International Conference on Automated Reasoning with Analytic Tableaux and Related Methods (TABLEAUX'09). Lecture Notes in Computer Science, vol. 5607, pp. 1–15. Springer, Berlin (2009). Invited talk
162. Jeavons, P., Cohen, D., Cooper, M.C.: Constraints, consistency and closure. Artif. Intell. **101**(1–2), 251–265 (1998)
163. Jeavons, P.G.: On the algebraic structure of combinatorial problems. Theor. Comput. Sci. **200**(1–2), 185–204 (1998)
164. Jeavons, P.G., Cohen, D.A., Gyssens, M.: A test for tractability. In: Proceedings of the 2nd International Conference on Constraint Programming (CP'96). Lecture Notes in Computer Science, vol. 1118, pp. 267–281. Springer, Berlin (1996)
165. Jeavons, P.G., Cohen, D.A., Gyssens, M.: Closure properties of constraints. J. ACM **44**(4), 527–548 (1997)
166. Jeavons, P.G., Cohen, D.A., Gyssens, M.: How to determine the expressive power of constraints. Constraints **4**(2), 113–131 (1999)
167. Jeavons, P.G., Cooper, M.C.: Tractable constraints on ordered domains. Artif. Intell. **79**(2), 327–339 (1995)
168. Jeavons, P.G., Živný, S.: Tractable valued constraints. In: Bordeaux, L., Hamadi, Y., Kohli, P., Mateescu, R. (eds.) Tractability: Practical Approaches to Hard Problems. Cambridge University Press, Cambridge (2012, to appear)
169. Jégou, P.: Decomposition of domains based on the micro-structure of finite constraint-satisfaction problems. In: Proceedings of the 11th National Conference on Artificial Intelligence (AAAI'93), pp. 731–736 (1993)
170. Jonsson, P.: Boolean constraint satisfaction: complexity results for optimization problems with arbitrary weights. Theor. Comput. Sci. **244**(1–2), 189–203 (2000)
171. Jonsson, P., Klasson, M., Krokhin, A.: The approximability of three-valued MAX CSP. SIAM J. Sci. Comput. **35**(6), 1329–1349 (2006)
172. Jonsson, P., Krokhin, A.: Maximum H-colourable subdigraphs and constraint optimization with arbitrary weights. J. Comput. Syst. Sci. **73**(5), 691–702 (2007)
173. Jonsson, P., Kuivinen, F., Nordh, G.: MAX ONES generalized to larger domains. SIAM J. Sci. Comput. **38**(1), 329–365 (2008)
174. Jonsson, P., Kuivinen, F., Thapper, J.: Min CSP on four elements: moving beyond submodularity. In: Proceedings of the 17th International Conference on Principles and Practice of Constraint Programming (CP'11). Lecture Notes in Computer Science, vol. 6876, pp. 438–453. Springer, Berlin (2011)
175. Jonsson, P., Nordh, G.: Introduction to the MAXIMUM SOLUTION problem. In: Complexity of Constraints. Lecture Notes in Computer Science, vol. 5250, pp. 255–282. Springer, Berlin (2008)
176. Jonsson, P., Nordh, G., Thapper, J.: The maximum solution problem on graphs. In: Proceedings of the 32nd International Symposium on Mathematical Foundations of Computer Science (MFCS'07). Lecture Notes in Computer Science, vol. 4708, pp. 228–239. Springer, Berlin (2007)
177. Jonsson, P., Thapper, J.: Approximability of the maximum solution problem for certain families of algebras. In: Proceedings of the 4th International Computer Science Symposium in Russia (CSR'09). Lecture Notes in Computer Science, vol. 5675, pp. 215–226. Springer, Berlin (2009)
178. Kahl, F., Strandmark, P.: Generalized roof duality for pseudo-Boolean optimization. In: Proceedings of the 13th IEEE International Conference on Computer Vision (ICCV'11), pp. 255–262. IEEE, New York (2011)

179. Khanna, S., Sudan, M., Trevisan, L., Williamson, D.: The approximability of constraint satisfaction problems. SIAM J. Sci. Comput. **30**(6), 1863–1920 (2001)
180. Khot, S.: On the unique games conjecture. In: Proceedings of the 25th Annual IEEE Conference on Computational Complexity (CCC'10), pp. 99–121. IEEE Computer Society, Los Alamitos (2010). Invited survey
181. Kohli, P., Kumar, M.P., Torr, P.H.S.: P3 & beyond: solving energies with higher order cliques. In: IEEE Computer Society Conference on Computer Vision and Pattern Recognition (CVPR'07). IEEE Computer Society, Los Alamitos (2007)
182. Kohli, P., Ladický, L., Torr, P.: Robust higher order potentials for enforcing label consistency. Int. J. Comput. Vis. **82**(3), 302–324 (2009)
183. Kolaitis, P.G., Vardi, M.Y.: Conjunctive-query containment and constraint satisfaction. J. Comput. Syst. Sci. **61**(2), 302–332 (2000)
184. Kolaitis, P.G., Vardi, M.Y.: A logical approach to constraint satisfaction. In: Creignou, N., Kolaitis, P.G., Vollmer, H. (eds.) Complexity of Constraints: An Overview of Current Research Themes. Lecture Notes in Computer Science, vol. 5250, pp. 125–155. Springer, Berlin (2008)
185. Kolmogorov, V.: Submodularity on a tree: unifying $l^\sharp$-convex and bisubmodular functions. In: Proceedings of the 36th International Symposium on Mathematical Foundations of Computer Science (MFCS'11). Lecture Notes in Computer Science, vol. 6907, pp. 400–411. Springer, Berlin (2011)
186. Kolmogorov, V.: Minimizing a sum of submodular functions. Discrete Appl. Math. **160**(15), 2246–2258 (2012)
187. Kolmogorov, V., Živný, S.: The complexity of conservative valued CSPs. Tech. rep. (2011). arXiv:1110.2809
188. Kolmogorov, V., Živný, S.: The complexity of conservative valued CSPs. In: Proceedings of the 23rd Annual ACM-SIAM Symposium on Discrete Algorithms (SODA'12), pp. 750–759. SIAM, Philadelphia (2012). Full version available on arXiv:1110.2809
189. Korte, B., Vygen, J.: Combinatorial Optimization. Algorithms and Combinatorics, vol. 21, 4th edn. Springer, Berlin (2007)
190. Krokhin, A., Jeavons, P., Jonsson, P.: Reasoning about temporal relations: the tractable subalgebras of Allen's interval algebra. J. ACM **50**(5), 591–640 (2003)
191. Krokhin, A., Larose, B.: Maximizing supermodular functions on product lattices, with application to maximum constraint satisfaction. SIAM J. Discrete Math. **22**(1), 312–328 (2008)
192. Kuivinen, F.: On the complexity of submodular function minimisation on diamonds. Discrete Optim. **8**(3), 459–477 (2011)
193. Kumar, T.K.S.: Simple randomized algorithms for tractable row and tree convex constraints. In: Proceedings of the 21st National Conference on AI (AAAI'06), pp. 74–79 (2006)
194. Kumar, T.K.S.: A framework for hybrid tractability results in Boolean weighted constraint satisfaction problems. In: Proceedings of the 14th International Conference on Principles and Practice of Constraint Programming (CP'08). Lecture Notes in Computer Science, vol. 5202, pp. 282–297. Springer, Berlin (2008)
195. Kun, G.: Constraints, MMSNP and expander structures. Tech. rep. (2007). arXiv:0706.1701
196. Kun, G., Nešetřil, J.: Forbidden lifts (NP and CSP for combinatorialists). Eur. J. Comb. **29**(4), 930–945 (2008)
197. Kun, G., O'Donnell, R., Tamaki, S., Yoshida, Y., Zhou, Y.: Linear programming, width-1 CSPs, and robust satisfaction. In: Proceedings of the 3rd Innovations in Theoretical Computer Science (ITCS'12), pp. 484–495. ACM, New York (2012)
198. Kun, G., Szegedy, M.: A new line of attack on the dichotomy conjecture. In: Proceedings of the 41st Annual ACM Symposium on Theory of Computing (STOC'09), pp. 725–734 (2009)
199. Ladner, R.E.: On the structure of polynomial time reducibility. J. ACM **22**(1), 155–171 (1975)
200. Lan, X., Roth, S., Huttenlocher, D.P., Black, M.J.: Efficient belief propagation with learned higher-order Markov random fields. In: Proceedings of the 9th European Conference on

Computer Vision (ECCV'06), Part II. Lecture Notes in Computer Science, vol. 3952, pp. 269–282. Springer, Berlin (2006)

201. Larose, B., Zádori, L.: Bounded width problems and algebras. Algebra Univers. **56**(3–4), 439–466 (2007)
202. Larrosa, J., Dechter, R.: On the dual representation of non-binary semiring-based CSPs. In: Workshop on Soft Constraints (CP'00) (2000)
203. Lauritzen, S.L.: Graphical Models. Oxford University Press, London (1996)
204. Lewis, H.R.: Renaming a set of clauses as a Horn set. J. ACM **25**(1), 134–135 (1978)
205. Lovász, L.: Submodular functions and convexity. In: Bachem, A., Grötschel, M., Korte, B. (eds.) Mathematical Programming—The State of the Art, pp. 235–257. Springer, Berlin (1983)
206. Mackworth, A., Freuder, E.: The complexity of constraint satisfaction revisited. Artif. Intell. **59**(1–2), 57–62 (1993)
207. Mackworth, A.K.: Consistency in networks of relations. Artif. Intell. **8**, 99–118 (1977)
208. Madelaine, F.R., Martin, B.: A tetrachotomy for positive first-order logic without equality. In: Proceedings of the 26th Annual IEEE Symposium on Logic in Computer Science (LICS'11), pp. 311–320. IEEE Computer Society, Los Alamitos (2011)
209. Madelaine, F.R., Martin, B.: The complexity of positive first-order logic without equality. ACM Trans. Comput. Log. **13**(1), 5 (2012)
210. Maróti, M., McKenzie, R.: Existence theorems for weakly symmetric operations. Algebra Univers. **59**(3–4), 463–489 (2008)
211. Martin, B.: First-order model checking problems parameterized by the model. In: Proceedings of the 4th Conference on Computability in Europe (CiE'08). Lecture Notes in Computer Science, vol. 5028, pp. 417–427. Springer, Berlin (2008)
212. Marx, D.: Approximating fractional hypertree width. ACM Trans. Algorithms **6**(2) (2010)
213. Marx, D.: Can you beat treewidth? Theory Comput. **6**(1), 85–112 (2010)
214. Marx, D.: Tractable hypergraph properties for constraint satisfaction and conjunctive queries. In: Proceedings of the 42nd ACM Symposium on Theory of Computing (STOC'10), pp. 735–744 (2010)
215. Marx, D.: Tractable structures for constraint satisfaction with truth tables. Theory Comput. Syst. **48**(3), 444–464 (2011)
216. Matiyasevič, Y.V.: Enumerable sets are Diophantine. Dokl. Akad. Nauk SSSR **191**(2), 279–282 (1970). In Russian; English Translation in: Sov. Math. Dok. **11**, 279–282 (1970)
217. McCormick, S.T., Fujishige, S.: Strongly polynomial and fully combinatorial algorithms for bisubmodular function minimization. Math. Program. **122**(1), 87–120 (2010)
218. Montanari, U.: Networks of constraints: fundamental properties and applications to picture processing. Inf. Sci. **7**, 95–132 (1974)
219. Motzkin, T., Raiffa, H., Thompson, G., Thrall, R.: The double description method. In: Kuhn, H.W., Tucker, A.W. (eds.) Contributions to the Theory of Games, vol. 2, pp. 51–73. Princeton University Press, Princeton (1953)
220. Mouelhi, A.E., Jégou, P., Terrioux, C., Zanuttini, B.: On the efficiency of backtracking algorithms for binary constraint satisfaction problems. In: Proceedings of the International Symposium on Artificial Intelligence and Mathematics (ISAIM'12) (2012)
221. Nagamochi, H., Ibaraki, T.: Computing edge-connectivity in multigraphs and capacitated graphs. SIAM J. Discrete Math. **5**(1), 54–66 (1992)
222. Narayanan, H.: Submodular Functions and Electrical Networks. North-Holland, Amsterdam (1997)
223. Nebel, B., Bürckert, H.J.: Reasoning about temporal relations: a maximal tractable subclass of Allen's interval algebra. J. ACM **42**(1), 43–66 (1995)
224. Nemhauser, G., Wolsey, L., Fisher, M.: An analysis of approximations for maximizing submodular set functions—I. Math. Program. **14**(1), 265–294 (1978)
225. Nemhauser, G.L., Wolsey, L.A.: Integer and Combinatorial Optimization. John Wiley & Sons, New York (1988)

226. Nešetřil, J., Siggers, M.H., Zádori, L.: A combinatorial constraint satisfaction problem dichotomy classification conjecture. Eur. J. Comb. **31**(1), 280–296 (2010)
227. Orlin, J.B.: A faster strongly polynomial time algorithm for submodular function minimization. Math. Program. **118**(2), 237–251 (2009)
228. Paget, R., Longstaff, I.D.: Texture synthesis via a noncausal nonparametric multiscale Markov random field. IEEE Trans. Image Process. **7**(6), 925–931 (1998)
229. Papadimitriou, C.: Computational Complexity. Addison-Wesley, Reading (1994)
230. Papadimitriou, C.H., Yannakakis, M.: Optimization, approximation, and complexity classes. J. Comput. Syst. Sci. **43**(3), 425–440 (1991)
231. Petit, T., Régin, J.C., Bessière, C.: Specific filtering algorithms for over-constrained problems. In: Principles and Practice of Constraint Programming (CP'01). Lecture Notes in Computer Science, vol. 2239, pp. 451–463. Springer, Berlin (2001)
232. Picard, J.C., Queyranne, M.: A network flow solution to some nonlinear 0–1 programming programs, with applications to graph theory. Networks **12**(2), 141–159 (1982)
233. Picard, J.C., Ratliff, H.: Minimum cuts and related problems. Networks **5**(4), 357–370 (1975)
234. Pöschel, R., Kalužnin, L.: Funktionen- und Relationenalgebren. DVW, Berlin (1979)
235. Post, E.: The Two-Valued Iterative Systems of Mathematical Logic. Annals of Mathematical Studies, vol. 5. Princeton University Press, Princeton (1941)
236. Promislow, S., Young, V.: Supermodular functions on finite lattices. Order **22**(4), 389–413 (2005)
237. Qi, L.: Directed submodularity, ditroids and directed submodular flows. Math. Program. **42**(1–3), 579–599 (1988)
238. Queyranne, M.: Minimising symmetric submodular functions. Math. Program. **82**(1–2), 3–12 (1998)
239. Raghavendra, P.: Optimal algorithms and inapproximability results for every CSP? In: Proceedings of the 40th Annual ACM Symposium on Theory of Computing (STOC'08), pp. 245–254 (2008)
240. Raghavendra, P.: Approximating NP-hard problems: efficient algorithms and their limits. Ph.D. thesis, University of Washington (2009)
241. Ramalingam, S., Kohli, P., Alahari, K., Torr, P.: Exact inference in multi-label CRFs with higher order cliques. In: IEEE Computer Society Conference on Computer Vision and Pattern Recognition (CVPR'08). IEEE Computer Society, Los Alamitos (2008)
242. Régin, J.C.: A filtering algorithm for constraints of difference in CSPs. In: Proceedings of the 12th National Conference on AI (AAAI'94), vol. 1, pp. 362–367 (1994)
243. Régin, J.C.: Generalized arc consistency for global cardinality constraint. In: Proceedings of the 13th National Conference on AI (AAAI'96), vol. 1, pp. 209–215 (1996)
244. Reingold, O.: Undirected connectivity in log-space. Journal of the ACM **55**(4) (2008)
245. Rhys, J.: A selection problem of shared fixed costs and network flows. Manag. Sci. **17**(3), 200–207 (1970)
246. Rose, D.J.: Triangulated graphs and the elimination process. J. Math. Anal. Appl. **32**(3), 597–609 (1970)
247. Rosenberg, I.: Reduction of bivalent maximization to the quadratic case. Cah. Cent. étud. Rech. Opér. **17**, 71–74 (1975)
248. Rossi, F., van Beek, P., Walsh, T. (eds.): The Handbook of Constraint Programming. Elsevier, Amsterdam (2006)
249. Roth, S., Black, M.J.: Fields of experts: a framework for learning image priors. In: IEEE Computer Society Conference on Computer Vision and Pattern Recognition (CVPR'05), pp. 860–867. IEEE Computer Society, Los Alamitos (2005)
250. Rother, C., Kohli, P., Feng, W., Jia, J.: Minimizing sparse higher order energy functions of discrete variables. In: IEEE Computer Society Conference on Computer Vision and Pattern Recognition (CVPR'09). IEEE Computer Society, Los Alamitos (2009)
251. Scarcello, F., Gottlob, G., Greco, G.: Uniform constraint satisfaction problems and database theory. In: Complexity of Constraints. Lecture Notes in Computer Science, vol. 5250,

pp. 156–195. Springer, Berlin (2008)

252. Schaefer, T.J.: The complexity of satisfiability problems. In: Proceedings of the 10th Annual ACM Symposium on Theory of Computing (STOC'78), pp. 216–226. ACM, New York (1978)

253. Schiex, T., Fargier, H., Verfaillie, G.: Valued constraint satisfaction problems: hard and easy problems. In: Proceedings of the 14th International Joint Conference on Artificial Intelligence (IJCAI'95), pp. 631–637 (1995)

254. Schlesinger, D.: Exact solution of permuted submodular MinSum problems. In: Proceedings of the 6th International Conference on Energy Minimization Methods in Computer Vision and Pattern Recognition (EMMCVPR'07). Lecture Notes in Computer Science, vol. 4679, pp. 28–38. Springer, Berlin (2007)

255. Schnoor, H., Schnoor, I.: New algebraic tools for constraint satisfaction. In: Complexity of Constraints, Dagstuhl Seminar Proceedings, vol. 06401 (2006)

256. Schrijver, A.: Theory of Linear and Integer Programming. Wiley, New York (1986)

257. Schrijver, A.: A combinatorial algorithm minimizing submodular functions in strongly polynomial time. J. Comb. Theory, Ser. B **80**(2), 346–355 (2000)

258. Schrijver, A.: Combinatorial Optimization: Polyhedra and Efficiency. Algorithms and Combinatorics, vol. 24. Springer, Berlin (2003)

259. Simeone, B., de Werra, D., Cochand, M.: Recognition of a class of unimodular functions. Discrete Appl. Math. **29**(2–3), 243–250 (1990)

260. Stoer, M., Wagner, F.: A simple min-cut algorithm. J. ACM **44**(4), 585–591 (1997)

261. Takhanov, R.: A dichotomy theorem for the general minimum cost homomorphism problem. In: Proceedings of the 27th International Symposium on Theoretical Aspects of Computer Science (STACS'10), pp. 657–668 (2010)

262. Takhanov, R.: Extensions of the minimum cost homomorphism problem. In: Proceedings of the 16th International Computing and Combinatorics Conference (COCOON'10). Lecture Notes in Computer Science, vol. 6196, pp. 328–337. Springer, Berlin (2010)

263. Tarjan, R.E., Yannakakis, M.: Simple linear-time algorithms to test chordality of graphs, test acyclicity of hypergraphs, and selectively reduce acyclic hypergraphs. SIAM J. Comput. **13**(3), 566–579 (1984)

264. Thapper, J., Živný, S.: The power of linear programming for valued CSPs. In: Proceedings of the 53rd Annual IEEE Symposium on Foundations of Computer Science (FOCS'12). IEEE, New York (2012). Preliminary version available on arXiv:1204.1079

265. Topkis, D.: Supermodularity and Complementarity. Princeton University Press, Princeton (1998)

266. Ullman, J.D.: Principles of Database and Knowledge-Base Systems, vols. 1 & 2. Computer Science Press, New York (1989)

267. Wainwright, M.J., Jordan, M.I.: Graphical models, exponential families, and variational inference. Found. Trends Mach. Learn. **1**(1–2), 1–305 (2008)

268. Werner, T.: A linear programming approach to Max-Sum problem: a review. IEEE Trans. Pattern Anal. Mach. Intell. **29**(7), 1165–1179 (2007)

269. Werner, T.: Revisiting the linear programming relaxation approach to Gibbs energy minimization and weighted constraint satisfaction. IEEE Trans. Pattern Anal. Mach. Intell. **32**(8), 1474–1488 (2010)

270. Willard, R.: Testing expressibility is hard. In: Proceedings of the 16th International Conference on Principles and Practice of Constraint Programming (CP'10). Lecture Notes in Computer Science, vol. 6308, pp. 9–23. Springer, Berlin (2010)

271. Zalesky, B.: Efficient determination of Gibbs estimators with submodular energy functions (2008). arXiv:math/0304041v1

272. Zanuttini, B., Živný, S.: A note on some collapse results of valued constraints. Inf. Process. Lett. **109**(11), 534–538 (2009)

273. Zhang, Y., Freuder, E.C.: Properties of tree convex constraints. Artif. Intell. **172**(12–13), 1605–1612 (2008)

274. Živný, S.: The complexity and expressive power of valued constraints. Ph.D. thesis, University of Oxford (2009)
275. Živný, S., Cohen, D.A., Jeavons, P.G.: The expressive power of binary submodular functions. Discrete Appl. Math. **157**(15), 3347–3358 (2009)
276. Živný, S., Cohen, D.A., Jeavons, P.G.: The expressive power of binary submodular functions. In: Proceedings of the 34th International Symposium on Mathematical Foundations of Computer Science (MFCS'09). Lecture Notes in Computer Science, vol. 5734, pp. 744–757. Springer, Berlin (2009)
277. Živný, S., Jeavons, P.G.: Classes of submodular constraints expressible by graph cuts. In: Proceedings of the 14th International Conference on Principles and Practice of Constraint Programming (CP'08). Lecture Notes in Computer Science, vol. 5202, pp. 112–127. Springer, Berlin (2008)
278. Živný, S., Jeavons, P.G.: Which submodular functions are expressible using binary submodular functions? Research Report cs-rr-08-08, Computing Laboratory, University of Oxford, Oxford, UK (2008)
279. Živný, S., Jeavons, P.G.: The complexity of valued constraint models. In: Proceedings of the 15th International Conference on Principles and Practice of Constraint Programming (CP'09). Lecture Notes in Computer Science, vol. 5732, pp. 833–841. Springer, Berlin (2009)
280. Živný, S., Jeavons, P.G.: Classes of submodular constraints expressible by graph cuts. Constraints **15**(3), 430–452 (2010)

Index

S. Živný, *The Complexity of Valued Constraint Satisfaction Problems*,
Cognitive Technologies, DOI 10.1007/978-3-642-33974-5,
© Springer-Verlag Berlin Heidelberg 2012